The Fifth "C"

The Fifth "C"

◆

The Criminal Use of Diamonds

Kelly Ross

iUniverse, Inc.
New York Lincoln Shanghai

The Fifth "C"
The Criminal Use of Diamonds

iUniverse books may be ordered through booksellers or by contacting:

iUniverse
2021 Pine Lake Road, Suite 100
Lincoln, NE 68512
www.iuniverse.com
1-800-Authors (1-800-288-4677)

ISBN: 978-0-595-46811-9 (pbk)
ISBN: 978-0-595-91101-1 (ebk)

Printed in the United States of America

For Carleen

Contents

Preface

Every day millions of dollars worth of diamond related criminal activity occurs in North America. Criminals are using diamonds and jewellery for money laundering, tax evasion, storing and moving proceeds of crime, frauds, and a multitude of other illegal activities. While this criminal activity has been going on in North America since the late 1800s, the general public has been victimized and few crime prevention measures have been employed. Recent statistics show that by value, these commodity types are among the fastest growing commodity sectors used by criminals. This is happening at a time when Canada has become the worlds' third largest producer of diamonds. In addition, law enforcement has focused little attention on diamond specific criminal activity, has few dedicated resources and virtually no legal tools to stop the activity. The diamond and jewellery industry have been proactive in attempting to limit the criminal exploitation of diamonds. For their part, they have drawn up guidelines that spell out what are acceptable and unacceptable business practices but these rules lack the legal bite that may otherwise provide a criminal deterrent.

The first step to preventing and truncating diamond related criminal activity is to gain an understanding of what is happening, why it is happening, and how it is happening. The diamond industry, gemmology, geology, law enforcement or intelligence practice each present one side of this complex issue. Knowing each side and how they are interwoven provides a more complete understanding of what is happening and why. This is the intent of this book; to provide law enforcement with an awareness of the scope and possibilities for the criminal exploitation of diamonds in Canada, North America and to lesser extent, other developed countries. Further, to stimulate discussion among law enforcement, industry, and stakeholders on what can be done to reduce the criminal use of diamonds.

This book details some of the criminal uses for diamonds, but this book is not all encompassing or absolute. Context is important to every situation and in that respect unless a specific case can be examined, concepts have to be generalized. There is also significant gemmological information contained in the book, but this is not a gemmology text and again ideas and concepts are generalized or simplified. I cannot reveal any police secrets or techniques and while I work for the

Royal Canadian Mounted Police (R.C.M.P.) this book in no way speaks for that organization. Any thoughts, opinions, or suggestions are my own or quoted accordingly.

I have had a strong interest in geology and later gemmology since I was about seven years old. This has grown into a life-long passion of which I have explored through the study of geology at the University level and gemmology through several schools. This drive has lead to a Bachelor of Arts degree focused on geology and geography, gemmology diploma's through two schools, and training as a rough diamond grader and valuator in both North America and Europe. My desire to further my knowledge and skills as a gemologist led to the 1999 launch of my internet-based diamond and jewellery business and more recently has led to consulting.

What I had not anticipated in opening my diamond business was what I would learn about the exploit potential of diamonds. Nor did I anticipate the usefulness of my industry knowledge and experience as a police officer. Yet, it is precisely this knowledge that I've acquired from beyond my police duties that in synergy with 18 years police experience, which is valuable in investigating the criminal use of diamonds. As a police officer I have been involved in such investigation across Canada and internationally and have provided expert testimony on the criminal use of diamonds and like commodities. Working in this field has also illuminated where knowledge gaps exist in law enforcement. This has fueled a desire to complete a Masters of Arts degree of Criminal Intelligence focused on the criminal use of diamonds and like commodities.

Several parts of this book may appear unflattering to law enforcement and the jewellery industry, but again in terms of the jewellery industry this is simply a reflection of the few that engage in criminal use of diamonds. This point cannot be understated. Several had suggested too write the book in such a manner that would give it a wider appeal to the general public. Others had suggested too include contact information about my diamond and jewellery business. However, the intention of this book is not to discredit the jewellery industry or law enforcement, or to generate jewellery business, but to raise awareness among law enforcement of the criminal use of diamonds.

There are so many people that had a hand in creating this book including people from industry and law enforcement that have been tremendously valuable in their support and contributions of knowledge, information, and/or guidance. Carleen, Mike, Lisa, Brent, Donna, Don, Stephen, Leah, John, Greg and the rest of you. Also a huge thank you to Rosie. Lastly, thank you to the Royal Canadian Mounted Police for allowing me to publish this book.

Introduction

There are a few things that I think it is important to keep in mind while reading this book. First, there is great exploit potential in the subjectivity of diamond valuations. In addition to this, the incredibly high values of diamond and its untraceable and undetectable qualities make it desirable for criminals to use.

Second, there is exploit potential in the lack of knowledge that North American's have about diamonds and the criminal use of diamonds and jewellery in general.

Third, when it comes to diamonds, not all exploitable opportunities exist in all countries. The conditions that may exist in the diamond industry, in government, law enforcement, or society in one country may not necessarily be transposed onto another country.

Last, and perhaps most importantly, I know many people in this industry and have met and dealt with countless others. Most are genuinely good people, the kind of people you would want for neighbors—honest, hardworking, and friendly. I highlight his point again because in reading this book, you may get the impression that the industry is fraught with criminals and this is not the case. Criminals like diamonds and this industry because there are excellent exploit opportunities, but rest assured, the diamond industry doesn't like criminals.

1

History of Diamonds

To understand the criminal desire to obtain and use diamonds, it is important to look back at the history of diamonds to see just how they have become so revered and valuable. Both the high value of diamonds and the associated criminal activity in acquiring these valuable items have been recorded throughout the ages. Criminal use of diamonds has been a common practice, however, in modern times the various methods of and reasons for the use of diamonds by criminals has changed. A quick look back through time shows how diamonds have worked their way into our culture and nearly all cultures around the globe, and symbolize strength, wealth, and power.

It is widely known that until the mid 1700's, diamonds came nearly exclusively from India. As a result, for other countries and cultures acquiring diamonds through India, the knowledge, fables, legends, and beliefs that the Indian culture attached to diamonds were passed on with the stones. There are writings from India that date back as early as 300 B.C.E. that explicitly talk about the value of diamonds. The text, known as the *Arthasastra*, was written as a manual specific to application of taxes and the title of the transcript is translated as *The Lessons of Profit*. This manual contains information on diamonds and gemstones. It outlines what makes them valuable, colour characteristics, shapes that were considered valuable and others that were not[1]. This text provides insight into the culture of India at this time and how highly diamonds were valued, so much so, that the export of diamonds and gemstones was prohibited. There are several other later day writings of Indian origin and other global locations. These include writings dated at the time of Christ from Greek authors. In fact the modern day word "diamond" is derived from the Greek word *adamas* meaning indomitable. One noteworthy Indian text, the *Ratna Sastras*, circa 400–1000 C.E., outlines the actual pricing scheme for diamonds and as Harlow[2] writes, there was an obvious market for diamonds at that time with only the best stones going to royalty. The overwhelming themes of the writing are of diamonds' value, and the beliefs about the powers that diamond possess. These beliefs include the thought that diamonds could cure illnesses, that diamonds were essentially a living creature that could reproduce, and that they held special powers that could protect warriors in battle.

Inclusive of this history are records of incredible daring thefts of diamonds, ancient and modern wars that were fought through diamond funding, and the remarkable pedigree of the most spectacular diamonds in the world. Through this ancient and modern history, diamonds have worked their way into popular culture as being an instrument of crime, a store of wealth, a symbol of luxury and decadence, and rightfully so. All this is true. Marilyn Monroe sang of diamonds; so does Madonna. Ian Fleming wrote several James Bond novels with plots that revolve around the criminal use of diamonds and jewellery, such as *Diamonds are Forever* and *Octopussy*. Of particular noteworthiness is the recent Bond film, *Die Another Day,* in which the criminal use of conflict diamonds not only captures a present day diamond issue but is key to the plot. Last, in James Cameron's *Titanic,* the giant blue gemstone that the main character throws overboard at the end of the show is assumed to be a diamond.

The weaving of diamond culture into present day life is simply a continuation of the historical reverence in which diamonds have always been held. From dozens of diamonds each with its own intriguing history, I have chosen a selection that captures the essence of how diamonds shape our society, stimulate our imaginations, and display the extremities of diamonds' value.

The Kohinoor or 'Mountain of Light'

The Kohinoor, pronounced 'Koh-i-Noor' or 'Mountain of Light' as it has been called, comes from India. The earliest history tells of this diamond weighing over 600 carats and belonging to the Rajah of Malawah[3]. The next account of this diamond's history is two hundred years later when the diamond becomes the property of Babur after he and his army defeated the much larger army of the Sultan of Delhi. The diamond, it is said, was valued beyond anything else of worldly origin. Several accounts of its value are represented beyond something that can be described by a dollar value. In fact the only way the Kohinoor could be valued was by giving it a value that could not actually be described monetarily. This was accomplished through comparing it with other nonsensical values that translated it into being worth more money than anyone would ever have. One description of the Kohinoor's value suggested the diamond was valued at half the daily expenses of the whole world[4]. A more recent owner of the Kohinoor, Nadir

Shaw, sold the diamond to a Sihk ruler Ranjit Singh c. 1813 to obtain protection. In attempting to place a value on the diamond, the diamond was sent to a diamond merchant who in turn told him the value could not be stated in monetary terms.

The diamond stayed in the possession of India royalty until colonization by the English in the mid 1800's. Under treaty the diamond became the property of the Queen of England but by this time its actual weight was 189 carats, having obviously been reworked over the years. The Kohinoor was shipped to England where again it was recut into a 108 carat oval brilliant which is presently mounted in the Queen Mothers crown. The full and complete history of the Kohinoor is rich with tales of how this diamond was used to finance war, obtain protection, and lost through war or war debts.

The Hope Diamond

Quite possibly the most famous and valuable of all diamonds, is presently stored at the Smithsonian Institute and is the most often viewed gem piece within the collection. This diamond originated in India and was brought to France in the mid 1600's with several other magnificent diamonds that were collected by Jean-Baptiste Tavernier. It was known at that time as the 'Great Blue' and was presented in its original 110 carat form to the French Monarchy who in turn had the diamond cut and reshaped into a 67 carat heart shaped wonder. This diamond was among several of the French Royal Jewels that is believed to have been stolen during the French Revolution. At this point the history is not particularly clear, however, it is likely the stolen diamond was smuggled to England and sold. In London it is widely believed but unconfirmed that the diamond was then cut again to its present form. This particular fact has sent many people on life time searches for the original 'Great Blue'. It is suggested that the recutting of the stone was a necessary evil, as the diamond would have been quite easily identified as the piece from the French Crown Jewels if it had not been modified. The result of recutting the diamond was a 45 carat antique cushion cut diamond, mounted in a pendant surrounded by several smaller white diamonds of various shapes. There are reports that two (2) other smaller blue diamonds were also obtained from the recutting process. One, was a six (6) carat stone apparently purchased by Karl II, the Duke of Brunswick, and Edwin Streeter apparently purchased the other, a one (1) carat blue in 1877. Today, the whereabouts of these two smaller diamonds is unknown.

The Hope Diamond was purchased for $90,000 in the early half of the 19th century by a fellow named Henry Hope, from which the present day name of the diamond is derived. The diamond stayed with the Hope family for nearly a century before being sold in some ambiguity c.1901. About ten years later, Pierre Cartier of jewellery fame came to be the owner of the diamond and shortly after the diamond was sold in 1911 to Mrs. Evelyn Walsh McLean. In 1949 the diamond was sold again, this time to Harry Winston of diamond and jewellery fame. Winston purchased the diamond for nearly $177,000. The Hope Diamond is presently held at the Smithsonian Institute after being donated to them by Winston in 1958.

This stone specifically plays into modern day folklore through James Cameron's movie *Titanic*. Most people are unaware of the history of the 'Great Blue' and this is not explained in the movie. However, the stone the main character throws overboard is a heart shaped stone like the original 'Great Blue' of the French Crown Jewels. This would suggest that the original French Blue was never re-cut after it was stolen, but was only lost or hidden away for some time, ultimately coming into the possession of the main character. She, of course, casts the blue stone overboard and never to be seen again. Based on this fictitious ending to the movie, the Hope Diamond would have come from some other source than the Great Blue. It certainly adds a twist and a specter of mystery to end the movie this way.

The Cullinan Diamonds

The Cullinan was named after Thomas Cullinan, the operator of the diamond mine where the Cullinan was found. This is the largest diamond ever recorded and weighed 3,106 carats or 1.37 pounds when it was pulled from a South African mine in 1905. To put this in perspective, the Cullinan is five (5) times as large as the 600 carat Kohinoor, the so-called 'Mountain of Light'. As the story goes, the diamond was actually found protruding from the wall of the mine and when extracted the mine manager said "… Mr. Cullinan will be pleased when he sees this!".[5] This is perhaps the understatement of the decade, if not the century, as nine (9) famous diamonds were eventually cut from this one stone. The largest of them is the Cullinan 1, a pear shaped, 550-carat beauty that was dubbed the Great Star of Africa, the largest gem quality finished diamond on record. This diamond is part of the British Royal Jewels and is mounted in the Royal Scepter. The Cullinan 2, known as the Lesser Star of Africa, is 317 carats and is mounted in the British Imperial State Crown. The other seven (7) Cullinan diamonds weighed approximately 94, 63, 18, 11, 8, 6, and 4 carats respectively and have all been acquired by the British Royal family.

The Golden Jubilee

In 1985, a large brown diamond weighing 755 carats was extracted from the Premier Mine of South Africa. This is the same mine that the Cullinan came from as well as other famous diamonds such as the 273 carat Centenary. The rough was believed to be of generally poor colour quality, and cutting it was more of a practice exercise. Surprisingly, the cut and polished diamond exceeded all expectations and resulted in a finished 546 carat fancy brown-yellow diamond, making it one of the largest gem quality faceted diamonds in the world. The Golden Jubilee was purchased from DeBeers in 1995 and given to King Rama IX of Thailand on the 50th anniversary of his coronation[6].

What is of importance, is the changing tastes of diamond purchasers. Whereas only the most colourless or colourful of diamonds were valued at one time, diamonds of deep brown-yellow colours are now also highly sought after. One has to wonder how many of these brown-yellow diamonds were discarded in the early days of diamond mining that would sell for high values in today's market.

Canadian Diamonds

The first three diamonds discussed in this chapter are of particular notoriety and have centuries of history. However, with diamonds having been found in Canada it is appropriate to discuss diamonds that originated in Canada's Arctic. Actually holding and closely examining one of the great diamonds of the world is something only a handful of people will ever do. In fact it is next to impossible to even get a good look at any great gemstone on display. This was my experience in viewing the Swedish Royal Jewels. I was in Stockholm with an R.C.M.P. hockey team for the World Police and Fire Games and while there I managed to squeeze in a visit to see the Royal heirlooms. The Swedish Royal Jewels are on display to the public at the Stockholm Royal Place and while historically interesting, and the tour a bargain, gemmologically speaking the jewels were miles away behind glass. Distance, low lighting, and a 'no cameras' rule made for anything more than a casual scan impossible. However, at least they were on display and employing strict security measures. One can't help wonder, though, whether they were genuine jewels or simulants. Regardless of location, the great jewels of the world that are actually on display to the public are usually several inches, if not feet, behind protective glass. This, combined with poor lighting is far beyond the distance required to really have a good look at the stones and jewellers' work. So for someone who will likely never actually hold or inspect those other diamonds, getting ahold of some magnificent diamonds is quite a thrill.

I once got a close look at two 10+ carat Canadian diamonds at the same time. They are quite magnificent as shown in the picture. The one closest to my fingertips is a 10+ carat D colour, flawless, pear shaped beauty. The other is also 10+ carat, but fancy canary yellow colour, eye clean and princess cut. These two diamonds could possibly fetch over $1 million. There was a third diamond and although difficult to see in the image, it is also noteworthy. It is the one in the case at the bottom right side of the image. This was a 13+ carat, fancy brown-yellow diamond, eye clean, and oval cut. This is also a sensational gem and like the Golden Jubilee, of a colour that is highly sought after in today's market.

American Diamonds

Few people have heard of the 'Crater of Diamonds State Park', in Arkansas, United States. However, this State Park created in 1972 may be the only place in the world where people can search for real diamonds. There have been over 75,000 diamonds found on what is now park property since the first diamond discovery in 1906, and many of these are quite remarkable.[7]

The Kahn Diamond

Natural fancy yellow diamonds are rare regardless of the country of origin. However, as chance would have it, a park visitor found a flawless 4.25 carat, fancy canary yellow diamond in Crater of Diamonds State Park. It was quickly sold to an Arkansas jeweller named Stan Kahn[8]. This diamond is by no means the most splendid of diamonds, however, this stone still commands a great deal of attention on its qualities alone—flawless, canary yellow, and 4.25 carats. The characteristics that can set one diamond apart from another are the historical characteristics that the diamond possesses. What a diamond needs to become famous is a unique history, or pedigree. This is a quality that the Kahn Diamond is well on its way too achieving. Since being found in 1977, this diamond has been fashioned three times into jewellery and has been worn by Senator Hillary Clinton. The last two times were for then President Clinton's inaugural ball in 1993[9] and again in 1997. With this, there is now the beginning of a history or pedigree for the diamond that could very well be the catalyst for other future events of significance for the diamond to gain further notoriety. At the very least, the connection of the Kahn Diamond to the Clintons and those events increases the diamonds value dramatically.

Some other honorable mentions from Crater of Diamond State Park are:[10]

- The Uncle Sam Diamond found in 1924, finished weight 12.42 carats.
- The Star of Arkansas found in 1956, finished weight 8.27 carats.
- The Amarillo Starlight found in 1975, finished weight 7.54 carats.

2

The Diamond Industry

The rough diamond industry has been in existence for centuries longer than the finished diamond segment of the industry. This is because the ability to cut and polish diamonds as we know it today really did not come into its own until the 17th century. Prior to this, diamonds that were used for adornment were fashioned into jewellery in their natural uncut state or with simple cuts showing brilliance or fire but without capturing either very well. At this point, the rough diamond business and essentially the retail trade of diamonds were one and the same. There was little industry stratification. However, diamonds were not a household item as they are today and were strictly reserved for royalty and the very wealthy. While the cutting and polishing industry was advancing, a new source of diamonds was discovered in Brazil in the 1700's. This opened the doors for the sale of diamonds to an even larger market. In the mid to late 1800's, the South African diamond finds vastly increased the availability of diamonds and the size of the industry grew. With this growth the finished diamond industry and the rough diamond industry began to traverse different paths. Today, the production of rough diamonds and ultimately the sale of finished diamond to the consumer are obviously connected but there are several links in the chain that separates the mine to market.

Diamond Production

The rough diamond industry as we know it today produces diamonds from several localities around the world. Every continent has the potential to have places where diamonds could be found. However, diamonds found on the different continents are mined in different ways, due to the deposit type. Each of these deposit types requires a different method of extracting the diamonds. In Canada, Russia, Australia, and much of South Africa, the diamonds are located in kimberlite and rarely in lamphorite. Kimberlite and lamphorite ore bodies are conical in shape like a carrot or pipe that extends deep into the earth's crust. They are the solidified rock remains of an ancient volcano-like deposit. These ore bodies are often called pipes. Scientists estimate that many of the kimberlite pipes originated at depths greater than 150 km below the earth's surface. In violent eruptions similar to an exploding volcano, molten rock cut through the layers of rock above it until it burst free at the earth's surface. In Canada the most recent of these types of eruptions are estimated to be between 50–70 million years old. Scientists are able to estimate the age of these eruptions through several means, usually involving the analysis of the age of the rock in the kimberlite. However, with some of the Canadian mine sites they are analyzing plant materials found deep inside the mine for clues of how old the deposit is.

This may seem odd, finding plant material mixed in with the rock several hundred feet below the earth's surface, but plant material, specifically ancient tree trunks, are being extracted from some of these mines. For the most part, these pieces of wood and tree trunks are perfectly preserved; not even fossilized after 50+ million years. There is an explanation for this. At the time these eruptions took place there were complete forests overgrowing the surface of the earth where the eruptions occurred. The eruptions were powerful but of short duration and when they breached the earth's surface they vaulted the trees, dirt, rock and molten rock together into the air like a tossed salad. When the eruption stopped, the material that was in the air above the blowhole fell back to the ground and filled in the hole. Many of the pieces of trees were buried over with several feet of rock. The heat of the eruption and dynamics of the eruption gases left the tree trunks in an environment that would keep them perfectly preserved over all this time. What is particularly interesting about this is that the species of tree found in the mine are of a type suited only for a climate much warmer than it is there today. This gives other scientists an insight into what the climate in the Northwest Territories used to be like. Similar ancient tree trunks have been found even further north on Axel Hieberg Island.

As the molten rock material broke its way to the surface during the eruption, it cut through several layers of rock. Some of these layers of rock contained diamond, because it is at these depths that diamonds are formed. Along the way to the surface, the molten material dragged with it pieces of rock from each layer that it cuts through and as such it brought the diamonds to the surface with it. Once the eruption was over the molten material, with all the other pieces of rock mixed in with it, cooled down and hardened in place. This matrix of rock material is what is known as kimberlite and within the kimberlite is boulders of rock that contain diamond. The diamonds may have been formed billions of years earlier and would have stayed in place if it were not for the eruption that brought them to the surface.

These mines are termed hard rock mines because the kimberlite ore body is hard rock material. Extraction of the kimberlite is done through strip mining techniques that exhaust surface deposits which then require extracting the rest of the diamond through building a mine shaft into the ground within or alongside the 'pipe'. In these mines the kimberlite rock is blasted and broken up into large chunks that are fed into a crusher, which pulverizes the rock into small particles that in turn releases the diamonds. Sometimes the crushing process damages the odd diamond crystal but most of the diamonds come away unscathed.

Open Pit Diamond Mine. Photo Courtesy of GIA[11]

The diamonds are separated from the crushed rock through different means. Some mines use an x-ray sorter, and others use a greased conveyor belt. For x-ray sorters, all the rock and diamonds that have been freed from the kimberlite rock are dropped past the x-ray beam one by one in a stream of pebbles and stones. This method exploits the fact that diamonds react in particular ways when exposed to x-rays. The x-ray beam is wired to a second component which is an air blaster located just below the x-ray beam. When the x-ray machine spots a diamond that has just gone through the beam, it sends a signal to the air blaster that is timed according to the rate of fall of the stones. When the diamond falls past the air blaster, a blast of air is emitted that sends the diamond into a collection container.

The greased belt method has the crushed kimberlite rock being dispersed over a greased conveyor belt. This method exploits the fact that diamonds are particularly attracted to grease and while most other rocks will not stick to the conveyor belt the diamonds will stick to it and will be held in place until they are scraped off. The diamonds are cleaned and bleached in each method and ready for the next process.

Hard Rock Mine Security

The security of these mine sites represent a controlled environment for producing diamonds where the mine owners know who their employees are and through due diligence, can create a workforce that screens out potential for theft or criminal activity. They know quite accurately how much their diamond production should be and can establish if diamonds are going missing during the process. The reason they are able to do this is two fold. First, they have knowledge of how many carats of diamond should be produced for every ton of rock that is crushed. For pure example sake, say on average there were 5 carats of diamonds per ton of rock crushed. Within the diamonds produced, there is a carat weight and quality distribution known as *run-of-mine*. This natural distribution of diamond weight and qualities means that in any mine there are going to be only so many 2 carat diamonds or larger, so many 1 carat diamonds, so many .50 carat diamonds and so on as well as a spectrum of diamond qualities both good and bad. Thus within a ton of rock there may be one 2 carat diamond, one 1 carat diamond, two 0.5 carat diamonds and the rest of the weight made up of smaller ones that on average make up 5 carats total weight. Perhaps in the next ton of kimberlite the yield is several small diamonds less than 0.50 carat that together weight approximately 5 carats. Perhaps in the next ton there is one large 5-carat diamond. Also with this distribution, there is a certain percentage of gem and near gem quality diamonds as well as industrial quality diamonds. For the average hard rock mine in South Africa, gem to near gem quality diamonds amounts to 30% of the diamond production, with the remaining 70% being industrial quality.[12] The distribution will change with each ton of rock but through this, geologists can establish the run of mine. Any significant thefts from the mine could show up as reduced carat weight per ton and or a noticeable change in the weight distribution from the norm. Production in hard rock mines is virtually hands off, aside from those responsible for taking the diamonds from the collection containers to the next step of the process, and there are tight security measures in place for each step. Tight security measures, employing good people, and best practices are benefits that the Canadian mines have been able to incorporate into their production models.

The Canadian mines have also had the luxury of examining the security models of other similar mines and creating a security model that excludes the pitfalls experienced elsewhere. One such pitfall happened in Australia at the Argyle mine in the 1990's where a high ranking mine security official was recruited by organized crime. In this case the mine official with the help of organized crime stole diamonds from the mine and moved them out of Australia and reportedly through a diamond dealer in Switzerland. From there the diamonds made their way to buyers in Hong Kong and Europe in what is estimated to be a $30 million loss to the Argyle mine[13].

One obvious security component that Canadian mines have is remote locations. This gives them an added measure that reduces the potential of criminal infiltration. On the other hand, there is a substantial increase in cost for developing a mine in the remote Canadian north. The mine site security in Canada is among the best in the world, if not taking top honors. It has the benefits of an enhanced security model, remote location, and some of the finest security personnel in the world including retired Royal Canadian Mounted Police officers and municipal police officers. Generally speaking, thefts from these operations are long term plans that seek to cash in after years of plotting and the greatest security arrangements and personnel won't stop it, not in South Africa or Australia, and not in Canada. Like the Argyle mine theft, it may very well be the result of a conspiracy among several people within the operation working together that defeats the security checks and balances. Information held by the Criminal Intelligence section of Canada makes note of the potential for organized criminal infiltration into the rough diamond sector of the Canadian diamond industry[14].

The second point to ponder regarding small time theft or opportunistic theft is that of high living standards in Canada compared with that of several other diamond producing countries. This high living standard significantly reduces the risk versus reward ratio involved in simple or opportunistic theft. For instance, to steal a large diamond in a country where the value of the diamond could significantly increase the person's personal wealth and well being may be enticing enough to risk getting caught. In Canada, the mineworkers make a handsome wage at the diamond mines and even a nice 2 carat rough diamond wouldn't noticeably enhance their personal wealth. They would risk losing their job, reputation, and way of life. On top of this, the ability to easily move or sell a stolen rough diamond in Canada is like trying to sell a pair of goalie skates in the Congo. For the average person, the market potential for rough diamonds in Canada is poor and one would need to be connected to the right people. Certainly, there is a market for rough diamond. It is very small, but it is growing.

Placer Deposits

Placer deposits are a concentrated accumulation of diamonds in a location that is not the location where they were brought to the surface of the earth. In this case, once having been freed by erosion from the kimberlite or lamporite pipe, they are transported to another location by gravity, wind, or water. Ultimately the diamonds may end up on a riverbank or sand bar, ocean floor, or beach. For riverbed deposits, the unsophisticated miner uses a basic sluicing technique similar to that used for gold panning. Placers deposits, beach, and ocean floor deposits represent a wide variety of localities where diamonds are found. Riverbed placer deposits are the primary source of diamonds in Brazil and are also common to West African and other South American diamond producing nations. These include, among other nations, the Democratic Republic of Congo, Sierra Leone, and Venezuela. In Brazil, diamond miners commonly use panning and sluice box techniques that are employed in riverbed gold mining operations. Diamond is approximately 33% heavier than quartz, which is the most common mineral on earth, and facilitates the use of gravity for panning for diamonds. These Brazilian miners are known as 'gampieros'[15]. They work along the rivers and using barges, move from one potential river site to another looking for diamonds. The working conditions are hard, the locations remote, there is virtually no local law enforcement presence, and there is a lawlessness there akin to the Wild West. In the Brazilian diamond industry, it is estimated that only 40% of the diamonds recovered are of industrial quality, somewhat less than the average of 65% industrial quality diamonds typically found in hard rock mines the world over[16].

In some countries, claims are staked for individual parts of the river's banks and the miners work only that part of the river. These deposits have security dynamics different than that of hard rock mines. First, it is difficult to protect the claim if it borders a river and unless one controls both sides of the river, there is one front that is uncontrollable. Many of the diamond deposits in West Africa are on rivers, which are essentially the borders to other countries. There is no run of mine to base production numbers and the quality of diamonds on, and as a result, theft is more difficult to determine through analysis of production. Many of the countries where riverbed deposits are exploited have little control over their industry, or reliable information on annual production numbers. In fact, it is the West African countries where diamonds illicitly mined or stolen from legitimate operations are being used to fuel civil conflicts and are so called *conflict diamonds* or *blood diamonds*. In addition, the opportunity for theft is significant as placer mining for diamond in many parts of the world is 100% hands-on. When a miner locates a diamond it is done so with his own hands and it is then turned over to the mine owner. Just as important as the mine security factor is the risk—reward ratio for opportunity theft. Again, in countries where death through war, famine, or disease is a daily event, the reward for stealing a diamond that brings significant wealth may outweigh the potential risks.

Non-river placer deposits are similar to river deposits in that they have been transported from the original kimberlite pipe to another location. They are usually associated to collections of materials at the base of an incline as a result of gravity's work or in ancient or dried riverbeds. Recovering diamonds from these deposits is a little more tedious as it requires sifting the material to expose diamonds and often requires a water source to aid in the separation process. This usually requires water being brought in from elsewhere and in doing so it creates additional mining difficulty and costs to the venture. One notable alluvial deposit for diamond is the Crater of Diamonds State Park in the United States and some African deposits. The U.S. deposit is noteworthy not because of the quality and volume of diamonds recovered from the deposit but because it is one of very few, small diamond deposits in the U.S. The Kahn Diamond mentioned in Chapter 1 comes from this deposit.

Ocean floor placer deposits are, as the namesake implies, located on the ocean floor, notably off the coast of West Africa. These diamond deposits are very rich in gem quality diamonds. Those off the coast of Namibia produce diamonds in the 90–95% gem quality range[17]. The diamonds have been transported by water, wind, or gravity over land and into rivers, which eventually empty into the sea. Techniques have been developed to extract the diamonds off the ocean floor using what best could be described as a huge shop vacuum aboard a large ocean vessel. This vacuum pump is directed by divers into crevasses and areas along the ocean floor that could have trapped diamonds or have a potential for diamonds. The vacuum picks up the diamonds, sea floor sediment, sea creatures, and whatever gets pulled into the vacuum. All materials picked up are drawn through the hose to the surface and onto a separating apparatus located on the vessel. The diamonds are extracted from the waste and collected for the next stage in processing. These mining operations have been moderately successful primarily because the quality of diamonds extracted is excellent. On the other hand, some of the issues that have plagued these ventures include finding reliable deposits, weather/water conditions, equipment failures, and ecological damage. Despite the relative isolation of a sea vessel, few workers, and a somewhat hands-free operation, million dollar thefts have occurred.

The beach deposits of Namibia are treasured for their high values and relative ease of extraction. These deposits are extensions of the ocean floor deposits but in this case, the diamonds that are emptied into the ocean are then washed up on sandbars or the shore. Over time, with wind and wave action, the relative high weight of diamond, and gravity, the diamonds work their way to the bottom of the sand on the beach, eventually coming to rest on the bedrock. Although diamonds are extremely hard, like anything else will break down over time through one of several processes of erosion. In this case mechanical erosion slowly desecrates those diamond crystals that have flaws, leaving only those with the purest of crystal structures—top gem quality stones. It is no coincidence that these diamonds have the strongest crystal structures. These diamonds have few, if any, small inclusions and are clear white in color, putting them into the highest end of the diamond value spectrum. Miners use heavy equipment to strip the sand from the beach until all that is left is a thin veneer of sand and any diamonds that are contained in the deposit. Low cost labour is then employed to comb the rock for diamonds. The diamonds are collected by the workers and passed on to the mine owners. This method of mining utilizes a high degree of hands-on labour, which again increases the loss potential through opportunistic theft and a local culture that sees no problem with stealing diamonds[18].

Theft from diamond mines is limited only by the imagination of criminals. Here is a sampling of some of the desperate and clever ways that criminals have stolen diamonds.

- In South Africa, pigeons were used by workers to take small bags of diamonds out of the mine site and directly to their home. This practice was uncovered by a mine site security officer who found a pigeon too bogged down to fly by a greedy worker who had strapped too much weight in diamonds to the bird. The result was an order to kill all pigeons on site.

- Mine site workers have concealed diamonds in personal hygiene products such as tubes of toothpaste or shampoo bottles.

- Desperate workers have gone to extremes in hiding diamonds under folds of skin, particularly the foreskin of the penis, or inserting them into the rectum. Some mines, particularly those in Africa where this is a problem, have subjected their employees to body and body cavity searches.

- Swallowing diamonds is a practice that has been extensively used to smuggle diamonds out of a mine. Again some mines have taken steps to identify perpetrators by making employees undergo an x-ray examination.

- Arrows that have had their core hollowed out and then filled with diamonds is another method that has been used to get the stones out of the mine. The arrows would be sent over the wall or fence to an accomplice waiting on the other side.

Sorting and Valuation

The next step in getting the diamonds to market is the sorting and valuation of the stones. Once diamonds have been extracted from their deposit source they have to be sorted, graded, and valued. The expertise required to sort diamonds is enormous as there are several thousand categories into which diamonds can be sorted.

The rough diamond market differs greatly from that of the finished diamond market in so many ways, not the least of which is terminology. With the finished diamond market, there are particular standards and names for colour, clarity, carat, and cut. These change from North America to Europe to Asia, but the standards and terms in each place can be married up with its counterpart grading system in another part of the world. With the rough diamond market, the terms and classifications (standards) vary from place to place, company to company, and country to country without a proverbial "Rosetta Stone" to tie the different languages of rough diamond together. Experience among those in the business is extremely important.

When referring to finished diamonds and weight it usually refers to carat weight. However, there are different ways that rough diamonds are referred to respecting their weight. When speaking of rough diamond under 0.66ct it is common to refer to the diamonds based on their sieve size. The diamond sieve size is established through running the diamonds through a series of sieves. The sieves have openings in them of a particular diameter and only diamonds of that size will fall through. Diamonds that are over 0.66 carat are more likely to be referred to by carat weight because these rough gemstones, once cut and polished, return larger sizes of diamonds where the actual carat weight is more critical to the value. In these larger +0.66ct sizes, the rough diamonds may also be referred to by grainer sizes. One grainer is approximately equivalent to 0.25 carat and therefore a four-grainer diamond would be approximately 1.00 carat. Grainer sizes are not precise and are more of a size class. For instance 0.66ct–0.85ct may be a three grainer and 0.86ct–1.10ct would be approximately four grainers. The following is typical Antwerp sieve sizes[19].

Antwerp Sieve Sizes

Size Class	Carat wgt. +/-	Size of Sieve Openings
-3	<0.02	-
+3	0.02–0.05	1.35 mm
+6	0.05–0.15	1.70 mm
+10	0.15–0.40	2.50 mm
+15	0.40–0.90	3.50 mm
+21	> 0.90	4.70 mm

Once the diamonds are separated into their different sieve or carat classes, they then have to be graded based on the clarity of the diamond. Rough diamonds contain inclusions, cracks, and imperfections, with the amount in each stone being different than the other. The more imperfections, the lower the quality and value of the gemstone. In any mine production, there are diamond qualities that are extremely good to extremely poor. A rough diamond with a large number of inclusions, those filled with cracks that permeate the entire stone, or those diamonds with crystal structures that grew poorly are unsuited for jewellery. Conversely, a well formed diamond crystal, clear and white, could produce a beautiful finished diamond or two. As such, rough diamond is classified into three main categories known as Gem/Near Gem Quality, Rejections, and Boart. The Gem/Near Gem Quality has several sub classes depending on shape and clarity. This is only one classification system and there are several different classification systems being used depending on the company involved or the country in which the rough is mined.

Gem/Near Gem—This category is made of diamonds that are easily cut into diamonds and relatively free of inclusions. Recovery weight could be as high as 60% and as low as 10% depending on the sub-category.

There are four main sub-categories based on the shape of the rough diamond:

1) Sawables—these diamonds are usually octahedral or dodecahedral shaped crystals and can typically be cut into two pieces that each can be faceted into

a finished diamond. Total recovery weight for the rough stone is usually between 40%–50%. A 1.00 carat rough diamond in this category may produce finished diamonds that together weigh approximately 0.40–0.50 carats. These shapes usually make up 50% of any rough diamond parcel.

Octahedral diamond crystal form

2) Makeables—these are irregular shaped diamonds or diamond pieces that are blocky. They can often be faceted 'as is' and may not require any preworking of the rough stone such as removing unwanted parts of the rough stone. These shapes will produce only one finished diamond and tend to have a fairly high recovery weight of 30% but in some cases it may be a high as 60%.

3) Maccles—these are diamonds with unusual crystal structure that makes cutting and polishing difficult. Gemmologically speaking, maccles are like Siamese twins—two diamonds that sprouted outward from the same point, each a mirror image of the other. They are typically, but not always, known as being triangular in shape and they can sometimes produce large finished diamonds. These diamonds are usually difficult to cut because in the middle of the stone is the twinning plane, the point that divides the two stones. The twinning plane is difficult to cut through and to polish because of the different orientation of the diamond's crystal structure on each side of the twinning plane. These diamonds have varying recovery weights.

4) Clivages—these are usually slivers of diamond pieces that may produce one or two small diamonds. Typically the recovery weight is very low. These diamonds are often referred to as Indian goods, making reference to the fact that stones like these are usually only cut and polished profitably in India where labour costs are low.

Rejections—These are the lowest quality of rough diamonds. However, they may contain small areas within the rough diamond that are suitable for polishing once it has been freed from the bulk of the poor quality diamond surrounding it.

Boart—This is the lowest quality of diamond with no gem value and is used for industrial purposes.

It has been suggested that with today's broad spectrum of jewellery fashions and demands for diamonds of all colours and clarities, rough diamonds may be better classified into cuttable rough diamonds (gem) and non-cuttable rough diamond (industrial)[20].

Interestingly, it is the industrial value of diamond for machining of metal and not the gem value that was a catalyst in the early attempts to produce man-made diamonds. In the 1940's the United States depended on a supply of diamond from South Africa for its machining purposes, particularly for the production of war materials. During World War II, with access to diamond disrupted and German U-boats running around the oceans, the supply of diamond to North America was reduced to critically low levels. After the war, the idea settled that producing man made diamonds or finding a North American deposit would prevent a similar future problem[21].

The colour of rough diamonds is much more difficult to determine than in finished stones where they can be compared to the colour of a master stone. As such the categories are rather broad and are typically named white, yellow and brown. Within each of these categories can be several sub-categories, each with a subtle different colour than the next. The number of categories is subject to the system being used to grade the colour. Still the standard 23 level grading scheme is regularly used for rough diamond. The 23 grades start from the letter D to Z with the letter D being the best or colourless and each following grade having progressively more yellowish or brownish colour.

Valuating a rough diamond is another matter altogether. The valuator has a head start from the diamond sorter because the sorting process has already begun to separate the rough stones based on a potential finished diamond weight. From this simple beginning, the valuator must examine the stone for the colour and clarity, and then determine the best possible shape of diamond to cut from the rough to maximize the stone's value. A long thin rough crystal may produce a higher weight if it is cut into an emerald or marquis shape rather than a round shape. Similarly a maccle may produce a higher recovery weight in cutting a triangle shaped diamond versus a round diamond. The possibilities and permutations are endless but if the valuator was to deal simply with round diamonds, they can establish what the final weight of the diamond will be and in turn the value of the rough stone. With the colour and clarity established, the valuator has to establish where the diamond should be cut to maximize the finished weight. The typical rough diamond octahedron crystal has three axes. The valuator measures the diameter of the rough diamond from these axes and can determine what the diameter of the finished diamond will be. Through this the valuator can establish what the approximate finished weight of the diamond cut from the rough diamond would be. For instance, a 5.1mm diameter of the rough diamond crystal would equate approximately to a 0.50 carat round cut diamond, a 4.1mm diameter rough diamond equals approximately a 0.25 carat round cut diamond, and so on. Marry this carat weight information up with the colour and clarity that has been established for the rough diamond and assume that this stone will have a good cut proportions, and you have established the value characteristic of the stone. These value characteristics are also known as the Four C's: Carat, Color, Clarity and Cut. Once the Four C's are established, a price on the rough stone while leaving room for the cost of cutting and a profit margin for selling the rough.

On a well formed octahedral crystal (a.k.a sawable) two diamonds can be cut from the rough diamond. This may be a 50/50 split of the rough or perhaps a 65/35 split of the stone. The 50/50 split requires cutting the rough crystal in two equal pieces and polishing a diamond from each half resulting in two finished diamonds of approximately the same weight. The 65/35 split is similar, but one of the pieces is larger than the other and as a result, there is one large finished diamond and one small finished diamond. The valuator then has to add the value of the two finished diamonds together in order to establish the value of the rough. Perhaps because of an internal feature, there may be only one diamond cut from the rough, and this is common. This could be the case if a large imperfection or serious structural flaw exists in one part of the stone. There is no use polishing up a second diamond if that piece is so heavily included as to be worth less than the cutting and polishing costs.

In looking at the North American diamond market, it is interesting to note that Canada is one of the largest producers of rough diamond and North America the largest consumer. Yet the process of sorting and valuating diamond is essentially done in Europe via Antwerp and London. This is basically what is done with Canadian rough diamonds. Once mined, Canadian rough diamonds undergo a very basic sorting procedure for the purpose of establishing royalties. The royalties are determined through an analysis of the rough diamond parcels and the Department of Indian and Northern Development (D.I.A.N.D.) administer this process. D.I.A.N.D., as it is known to the industry, does not have anyone on staff that can place a value on the rough diamonds, and this is contracted out to the private sector. At approximate intervals of 3–5 weeks, diamond valuation experts from Europe under contract from D.I.A.N.D., come to Canada to evaluate the parcels of diamonds.

Another government department, Natural Resources Canada (N.R.Can) is responsible for issuing Kimberley Process (K.P.) certificates that certifies the Canadian diamonds are 'non-conflict' diamonds. The Kimberley Process is a United Nations backed endeavor to eliminate the trade of 'conflict' diamonds. So called, 'conflict' diamonds are so named because they have been used by West African groups at war, to purchase arms to support their conflict. All rough diamonds exported from Canada require this certification in compliance with the Export and Import of Rough Diamonds Act. Once government valuation and Kimberley Process certification is complete, the diamonds are sent to Europe to be more thoroughly valued and sorted. Approximately 10% of gem quality rough diamonds mined in Canada end up back in Canada to be cut and polished; the rest are sold on rough diamond markets in Antwerp and London. This is very much a function of history but also a reality of where the expertise is located.

There is a potential for Canada to expand on its relatively small rough diamond market or for the establishment of more cutting facilities. One hurdle is the lack of trained personnel to work in diamond sorting, valuation, or cutting facilities. There are few formal training centers to teach people these skills and like most diamond knowledge and skills, it only becomes sharp through experience. This lack of skilled personnel is evident in the government contracting of experts from Great Britain for the purpose of assessing royalties on rough diamond parcels leaving Canada. In addition, the majority of diamond cutters in Yellowknife are immigrants from other diamond cutting centers of the world. There are few rough diamond dealers in Canada and no proclaimed center of rough diamond dealing or open diamond bourse (market) as found in Antwerp. Like sorting and valuating, there are few people in Canada with the expertise to profitably handle rough diamond. The other issue holding up a Canadian rough dealing center is the lack of a market for rough diamonds in Canada, because the main cutting and polishing facilities in Canada already get some or all of their diamonds directly through the mine producers. The chance of new markets for rough diamonds opening up in Canada in the shape of new cutting facilities is slim if they do not have an already established market to supply them with rough diamonds. It's a bit like the chicken and the egg conundrum—without a market to sell the rough diamonds, there is little chance of diamond dealer getting established and on the flip side, without established diamond dealers to supply rough, there is little chance of diamond cutting facilities being established. Having a steady supply of rough diamonds is crucial to having a profitable diamond cutting business.

Rough Diamond Dealing

Once the diamonds have been sorted and valued they are ready to be sold to the rough diamond dealers. The bulk of rough diamonds are sold at a diamond bourse. The diamond bourse is a rough diamond market of sorts where buyers and sellers meet to do business. There are a few rough diamond bourses, none more notable than the rough diamond sales arm of DeBeers known as the Central Selling Organization (C.S.O.). For decades, DeBeers has sold its rough diamonds through the C.S.O. with great success. The bourse is not open to the public in that it is limited to approximately 150 buyers from around the world, also known 'site holders'.

Approximately once per month DeBeers holds its rough diamond sales, which amounts to the site holders being presented with a box of rough diamonds that may or may not conform to a customer's wish list. The site holder has two options. One is to return the box hoping to get a better supply at the next sale, or purchase the box at the offered price. More often than not, the box is purchased at the specified price because if it is returned, the buyer is left without a steady supply of rough diamonds for their factories and with few options. Historically, a site holder who returns the box on consecutive occasions without making a purchase could lose the site at the diamond bourse[22]. The site holder may himself sell off some of the diamonds that were purchased from the C.S.O. to rough diamond buyers that do not have a steady supply of rough diamonds. In the past 30 years, diamond producers from Russia, Australia, and now Canada have chosen to sell their diamonds through alternative venues than the DeBeers. This has resulted in the emergence of new markets for rough diamond.

Finished diamonds

The first step in the finished diamond market is to cut and polish the diamond. Cutting the rough diamond removes the unwanted exterior part of the rough gem, bringing out the beauty of diamond so that it is prettier, more desirable, and easily mounted into jewellery. Those who cut and polish diamonds are craftsman with excellent skills and possess an intimate knowledge of the entire industry. Cutting a diamond is not an easy process as diamonds are the hardest natural substance, several times harder than the next hardest substance—sapphire[23]. In addition there are several steps in the cutting and polishing process, each requiring specialized training.

There are several steps that a rough diamond is put through to becoming a finished diamond. First the diamond is *bruted*. This is the process of rounding the diamond. Historically, bruting a diamond is done by grinding two diamonds together in a circular motion that ultimately rounds each one of the diamonds off to the desired external shape. In recent years, lasers have been used to round the diamond and create the shape. Next, the rough diamond is *roughed out*. This process is a cursory step that removes bulky unwanted surface material and orientates the rough diamond with preliminary cuts of the main facets. The next step is to *block* the rough diamond. In the first stage, this process polishes in four of the eight main facets (aka bezel facets) and four corresponding pavilion facets. In the second stage of blocking, the other four main facets and pavilion facets are polished. The final stage is the *brilliantering*. In this stage the rest of the crown and pavilion facets are polished and the diamond is complete[24].

Typical stages for cutting and polishing a sawable rough diamond into a finished diamond

Diamond facet being polished. Photo Courtesy of GIA[25]

Cutting diamond is difficult, however, diamond has a weakness that makes the diamond cutter's job easier. The bonding of carbon atoms in a diamond's crystal structure is weaker in some directions than in others. This is known as 'directional dependant hardness' and the result is that the weak spot of the diamond can be cut with the hard part of another diamond. If diamond is crushed up into tiny microcrystalline pieces, and this powder is used together with a cutting wheel, it can cut through diamond. The cutting wheel is specifically placed to cut through a part of the diamond that has the lowest directional hardness. With all the tiny pieces of diamond powder oriented in different directions on the cutting wheel, some of the hard pieces will cut through the softer part of the diamond. In time, the cutting wheel spinning very quickly cuts through the diamond. This is no quick process and can take several hours to several days depending on the diamond. Lasers are now being used to cut diamond but in only certain cutting centers and by limited manufacturers in Europe, Asia, and North America. This is largely a function of the material that is being cut in these centers and the potential profit margins that can be realized. The laser equipment is expensive and in order to be profitable, the manufacturers that use lasers need to show higher profit margins. The way manufacturers do this is to only cut larger, quality diamonds. New York, Antwerp, and Yellowknife are cutting centers that cut and polish only high end product and lasers are widely incorporated into diamond manufacturing. However, even India, a center that has historically cut low quality diamonds, has manufacturers that are now utilizing lasers in their factories.

In most of Europe and North America, the cost of cutting and polishing diamonds is much higher than other parts of the world such as Israel, India, or China. As a result, they usually cannot profitably produce cut diamonds smaller than 0.18ct. One of the ways to increase the productivity is to employ the use of robotics to polish the diamonds rather than the age-old manual polishing technique. Robotics operations can polish several diamonds at a time on one polishing wheel and do so in a fraction of the time it takes a person. Robotic polishers also have the ability to run 24 hours per day. The start up costs for robotics is of course higher but the pay off is immediate in terms of output of finished diamonds. In centers with high cutting cost, factories will tend to cut larger carat, high-end goods with margins that are fat enough to cover these cost and provide a profit.

Even at this level, the market is still very competitive and some companies have not been able to eke out a profit. Denton Cho diamonds and Sirius diamonds, both located in Yellowknife, were unable to make ends meet and found themselves in a position of bankruptcy or serious financial difficulty. Both companies have since found new sponsors to back them financially and have now continued operations. In Asia and India, where labour costs are significantly lower, it is financially viable to cut and polish small diamonds. For instance, in Yellowknife, Canada, the cost to cut a 1 carat diamond is approximately $250 where in India the cost to cut the same diamond is approximately $30.

World Diamond Cutting Centers

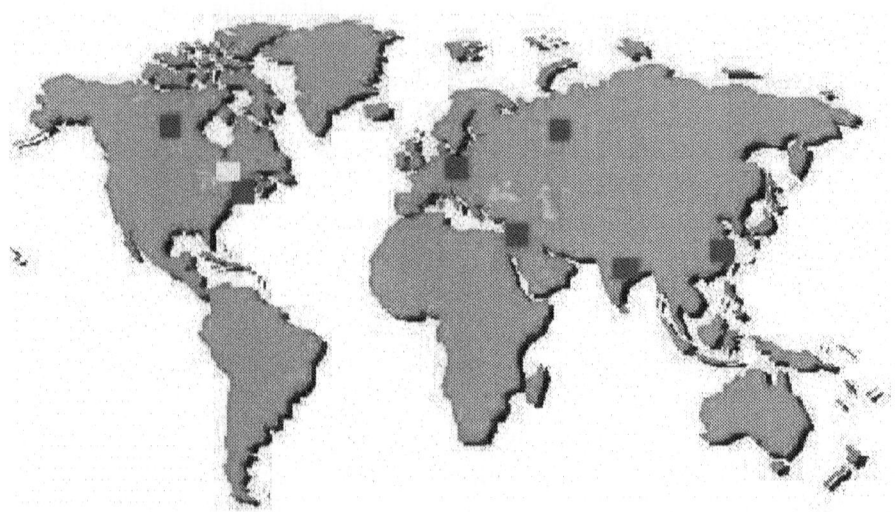

Left to Right: Yellowknife (Canada), New York (U.S.A), Antwerp (Belgium), Israel and Armenia, India, Asia. The light grey square illustrates the potential cutting centers of Toronto or Montreal (Canada)

Once diamonds have been cut and polished they are ready to be sold to finished diamond dealers. There is no change to the actual diamond at this end of the market except that the finished diamonds are sold at a profit to each successive link on the chain from here to the consumer. Diamond dealers typically buy the diamonds from the diamond cutters. These vendors deal solely in diamond, making large volume purchases directly from the diamond cutters. The largest buying group in the North American diamond-dealing world is the New York City Diamond Dealers Club. This group of diamond dealers supplies approximately 80% of the finished diamonds to the North American diamond market[26]. Diamond dealers then provide the diamonds to diamond wholesalers and jewellery manufacturers at a small margin. The wholesalers provide the diamonds to retail jewellery stores again at a small margin and the jewellery stores sell them to the consumers at higher margins again.

Although this illustrates the usual movements of diamonds from mine to market there are new developments in the industry that are seeing diamonds take a shorter route. Companies like BHP Billiton (BHP) and Aber Resources, who mine diamonds in Canada, are processing and cutting their own stones and selling them to the wholesale market or consumer under their own brand name. BHP, which owns the Ekati mine in the Northwest Territories, has seen a strong demand for their own 'Canadamark' branded diamonds. In this way, BHP is realizing a greater profit margin on their own brand of diamonds by skipping some of the traditional stages on the mine to market. Aber resources, a partner in the Diavik Mine, purchased Harry Winston Jewelers in a profitable move to bring their own high-end diamonds directly to the consumer. Tiffany & Co. of New York, the famous jewellery retailer, has sought to reap the same rewards as Aber and BHP through jumping several of the mine to market queues that exist for most of the diamond trade. Tiffany & Co. has signed an agreement to purchase $50 million U.S. worth of rough diamonds directly from the Aber Resources share of the Diavic mine output[27]. With this, Tiffany & Co. has set up the state of the art, Laurelton Diamonds cutting facility in Yellowknife. The results in all three ventures is a move away from the traditional multi-tiered European way of moving diamonds from the mine to market and to a more profitable approach that eliminates many of the middle men. There is another advantage to this approach in that these companies have a much greater control over where their diamond stock has come from and the movement of the diamonds from the mine to the consumer. This is particularly relevant when considering the issues of 'conflict' diamonds and illicit diamonds discussed in detail in Chapters 5 and 6. Perhaps this trend will be the wave of the future for the North American diamond markets.

3

The Value of Diamonds

From the beginning, diamonds were so revered that their value was measured far beyond anything financial in nature. In fact, at first, diamond's value was measured in other ways, according to beliefs of how it shaped the environment around us, such as its ability to ward off evil, heal the sick, or provide protection to those who wore diamond into battle. In time, these valuations were later replaced with those of a monetary nature. However, these too were lofty and in the beginning only the wealthiest of people could afford a diamond. It wasn't until the 20th century that the vast quantities of diamonds mined in South Africa came to the mass market. Suddenly diamonds were available to the entire world. The value of diamond has not diminished over the decades and for very good reason—a company named DeBeers. This company has held a virtual monopoly on the diamond industry for approximately 100 years and at one time controlled nearly 80% of the world's rough diamonds[28]. Although this company started out under the control of different people and with different names, the history is there to trace its roots back to the 1800's and the vision of Ernest Oppenheimer.

At the beginning DeBeers made a practice of gobbling up competitors and purchasing most of the rough diamonds from mines they did not own. Those who dared venture into the industry found themselves under enormous pressure. Many companies sold out to or were bought out by DeBeers and along the way Debeers managed to maintain control of the diamond market. Holding all the diamonds meant that Debeers could release whatever quantity or quality of their rough diamond stock to the market and in doing so, maintain the price levels and reduce oversupplies of particular sizes. This monopoly was held by DeBeers until a few decades ago, but this monopoly has created a stable market for diamonds while increasing demand year after year. This has been nothing less than lucrative for the industry as a whole. If this were not the case, then everyone would have gotten out of the business decades ago and left Debeers alone to bring diamonds from the mine to market.

On each step from mine to market there is a value added to the diamond, as every person takes a profit from moving the diamond towards the consumer. Every diamond is different and it is the grading standards of the Four C's that are used to put a specific value to an individual diamond. As mentioned previously, the Four C's are specifically known as Carat weight, Colour, Clarity, and Cut, with the quality of each one of these valuators affecting the price of the diamond. The higher the carat weight and the better the clarity, colour, and cut of a diamond, the more beautiful and rarer diamond is and ultimately a higher value.

The idea of the Four C's is both simple and complex. It is simple in that it is easy to understand that the higher the quality of each of these four valuators, translates into a more rare, beautiful and therefore valuable diamond. It is complex in how the quality or grade of each of the four valuators is established. There is really so much information required to fully understand the diamond grading systems created by various gemmological authorities around the world. As such, and for simplicity sake I have provided a condensed version of each of Carat, Clarity, Colour, and Cut.

Little will be discussed about rough diamond valuation except to say the process is similar to grading finished diamonds but the criterions are carat, colour, clarity, and shape (rather than cut). There is particular training and expertise that is altogether different for valuating rough diamonds than finished diamonds. For one thing, the shape of the rough diamonds can be nice crystals that can yield two finished diamonds or be a sliver of a diamond that may produce a small finished diamond only a fraction of the original rough diamond weight. The rough diamond may have a 'skin' on it that doesn't allow easy viewing of the interior to determine its clarity. Perhaps the most difficult part is establishing what the weight of the finished diamond that is cut from the rough diamond will be. This is extremely important because the value of the rough diamond is a reflection of the value of the finished diamond(s) that can be attained from the rough diamond. Those who work with rough diamonds in this capacity are relatively few and are among the most skilled of the diamond professionals. The grading practice has been developed over several decades through the work of diamond institutions like the GIA (Gemological Institute of America) and the HRD (The Antwerp Diamond High Council). Each institution has developed its own grading system for the Four C's and for the most part they are similar with only slight differences in terminology or standards. Among the few most experienced professional diamond graders, there is often little difference between a diamond graded by one gemmologist to another and the grading can be accurate to plus or minus one colour or clarity grade. Yet, even if the grades are close, the fact is that the grade given to a diamond by different gemmologists will almost always be different. Having said that, the reality is that the grading of colour, clarity, and cut are subjective and this means that there is no definitive and universally accepted way of declaring the precise quality of the diamond. The reason for this is that each of these particular qualities is left to the diamond grader to interpret. Thus, diamond grading is subjective rather than absolute. This subjectivity of diamond grading holds great exploit potential for criminals.

In terms of commerce, diamonds are bought and sold by the carat. As such, a diamond of a particular carat weight, for example 0.45 carat, that has particular quality characteristic in terms of a clarity grade, colour grade, and cut grade, will sell for a certain price per carat. For example, lets say $1,200 is the price per carat for this diamond. Therefore the cost of this diamond would be $540 based on multiplying the weight of the diamond (.45) by the price per carat for a diamond of those particular qualities ($1,200/ct).

$$0.45\text{ct diamond} \quad X \quad \$1,200 \text{ per carat} \quad = \quad \$540.$$

The trick to valuating diamond is knowing what the price per carat is of a diamond of any particular quality. Having this knowledge makes it quite easy to figure out the price of any diamond when there is some sort of pricing baseline for diamonds of a particular carat weight, clarity, colour, and cut. As it happens, there has been such a diamond-pricing baseline that is available to the industry since the late 1970's. The Rapaport or Rap, as it is known in the industry, is published on a continual basis in a paper format or can be downloaded off the Internet. The report is based on the high cash wholesale price of diamonds. Through this report, jewellers, wholesalers, diamond dealers, or cutters have a universal baseline from which to establish a price for the diamonds they are selling. The report is arranged in a series of tables that classifies diamonds into weight categories. Each tables correspond to a different weight class of diamonds, then within each table all the colour and clarity grades can be cross referenced which leads to a stated price per carat in U.S. dollars. Once the price per carat of the diamond is established the weight of the diamond can be multiplied by the price per carat to determine the wholesale price of the stone. The prices on the Rapaport assume the diamond is a good cut and recommend adding a percentage to the price if the cut is exceptional or subtracting a percentage for poorer cuts.

Below is an example similar to what a Rapaport report outlines however the values I have used in the table are not 100% accurate. I have only included a few supposed tables here but current values can be obtained from the industry. The table outlines the type of cut and the carat weight class. On the left side of the table are the colour grades and across the top of the table is the clarity grade. Assuming the diamond being examined has a good cut, then all of the Four C's are represented in the table. With that the high cash wholesale value of the diamond can be determined. For example, a 0.95 carat round brilliant cut diamond with good cut proportions of SI1 clarity, H colour, first go to the table that contains the values for the weight of the diamond. This would be the table for diamonds in the 0.90–0.99 carat weight class. Then find the colour on the left side and the clarity value across the top, and where the row and column intersect is found the price per carat of the diamond. In checking the table for H colour, SI1 clarity you find a value of 46. Multiply this value by $100 and the result, $4,600. is the price per carat for that quality of diamond. Now multiply the carat weight of the stone (0.95 carat) by the price per carat ($4600.), to determine the price of your stone.

$$0.95 \text{ carat } \times \ \$4,600 \ = \ \$4370.$$

Now this is price is in U.S. dollars and does not include the import taxes of GST. We will go over taxes later.

Round Diamonds 0.89–0.99 carat

	IF	VVS1	VVS2	VS1	VS2	SI1	SI2	I1	I2	I3
D	100	80	75	69	63	60	53	33	23	13
E	80	75	69	63	60	57	51	32	22	12
F	75	71	65	60	58	54	49	31	21	12
G	70	64	61	57	55	51	46	30	20	11
H	64	61	54	54	52	46	41	28	19	11
I	54	52	48	46	44	41	36	26	18	10
J	45	43	41	39	38	36	30	24	17	10
K	38	36	35	33	32	30	27	20	16	9
L	33	31	30	29	28	27	25	19	15	8
M	29	28	27	26	25	25	23	18	14	8

Round Diamonds 1.00–1.25 carat

	IF	VVS1	VVS2	VS1	VS2	SI1	SI2	I1	I2	I3
D	170	110	100	88	79	70	61	41	28	17
E	113	100	87	82	75	67	58	39	26	15
F	100	89	82	78	73	62	55	37	25	14
G	87	81	78	72	67	59	52	36	24	13
H	71	69	65	62	59	54	50	34	23	12
I	60	58	55	52	49	46	42	30	22	11
J	50	49	48	46	44	41	38	27	20	11
K	46	45	44	41	40	38	34	26	18	10
L	42	40	39	38	36	35	31	24	16	9
M	35	33	32	31	29	28	26	20	15	8

The Four C's

Carat Weight

Quite simply, diamonds become increasingly rare as their carat weight increases, therefore the greater the carat weight of a diamond the higher the price per carat. The weight of diamond is accurately determined with a carat scale to 0.01 of a carat or better. In real terms, carat weight can be equated as follows. 1 carat equals 0.2 grams, 5 carats equals 1.0 gram, 155 carats equals approximately 1 troy ounce, and there are 31 grams in a troy ounce. The weight of diamond is also described as points. The points are not facets or corners on the diamond but are a fractional or decimal representation of the carat weight with each point representing $1/100^{th}$ of a carat. For instance, a half carat diamond (0.50 carat) could be described as a 50 point diamond. Diamonds that are over 1.00 carat weight are usually referred to in carat weight while diamonds under 1.00 carat are described in points. Because the carat weights of diamonds are not uniform and the large diamonds are increasingly rare, the value of the diamonds versus carat weight is a non-linear scale. Two 0.25 point diamonds are not worth as much as one 0.50 point diamond, all things remaining equal (cut, clarity, colour). Similarly, two 0.50 point diamonds are not worth as much as a 1.00 carat diamond, again with all else remaining equal. The price per carat for diamonds of a particular quality jumps at regular intervals reflecting the increasing rarity of the diamonds. Whole-sale pricing reflects the non-linear pricing of diamond by categorizing diamonds into size classifications with per carat price jumps following each increase in size class. An example of size classes could be diamonds from 0.22–0.29 carat, 0.30–

0.39 carat, 0.40–0.49 carat, and so on. What is interesting to note is that although the price per carat jumps at each size class, the jump from 0.40–0.49 carat class to 0.50–0.59 class is particularly large as is the price jump from the 0.90–0.99 carat class to the 1.00–1.49 carat class. These diamonds at the half carat and one carat level are like benchmarks of which there is higher demand and increasingly lower supply. The higher demand is reflected in the larger price jump from the size class just below it. A simple pricing chart for the size classes of 1.00–1.49 carat and 0.90–0.99 carat was displayed in the previous pages. This is similar to the pricing charts used by the industry to establish the high cash wholesale price for a diamond they have.

For an exercise to help clarify this, use the chart to try to figure out the value of a 1.01 carat round brilliant cut diamond, of SI2 clarity and H color with good cut proportions. The answer can be found in the endnotes[29].

Colour

Colour of a diamond is based on the amount of colour that is visible in a diamond. Colourless diamonds are valued the most unless it is a fancy colour diamond which will be discussed later. The colour grades start at D, the most colourless of diamond grades, and progresses with the addition of yellow, brown, or grey tints to the grade of Z. The colour grade and in turn the quality, decreases as more colour tints of yellow or brown are observed. Therefore, colourless diamonds equate to higher per carat values. The exception to this is fancy colour diamonds, which are diamonds that have very strong colour. Such fancy diamonds have been found in the colours of yellow, red, green, pink, blue, and various other colours. Diamonds with these colours are extremely rare and as such usually command the highest per carat values.

Colour is also seen from diamonds in rainbow flashes that can be seen emanating from the stone in what is known as dispersion. This is not the body colour of a diamond but will be mentioned here anyway. Few precious gemstones have this property of dispersion and this is one of the qualities that are valued in a diamond. What happens to some light that enters and then exits a diamond is similar to that of a prism. The light that exits the diamond is spread out into its spectral colours. Although the rainbow colour flashes is not colour from the body of the diamond but light that is seen as colour, it is actually brought out as a result of the cut, which will be discussed last.

The colour grading of diamonds, as previously mentioned, carries with it a degree of subjectivity. The reason for the subjectivity is simple. When grading a diamond's colour, the gemmologist uses master stone sets to compare the diamond with, and through this, assigns a colour grade to the stone. These master stone sets are usually made of cubic zirconium that is coloured specifically to match the colour of diamond master stones. The master stone sets are usually made of five stone colours of D-F-H-I-J or E-G-I-K-M. Complete 10 stone sets of all the top colour grades are also available. Gemmologist that need to establish the colour grade of a diamond use full spectrum light, which is a light source that mimics the same light spectrum as the sun. On a white background the diamond is placed side by side with the master stones and matched to the master stone that appears to be the closest in colour. This is best done by inverting the diamond and the master stone in a white colour grading tray so that the gemmologist is actually looking for colour in the bottom (pavilion) of both the diamond and the master stone. Gemmologists that use a complete set of master stones can easily and with a degree of certainty establish the colour grade for the diamond they are grading. However, each person sees colour differently and it is very possible for one person to see the colour of a stone differently than another. Also when the diamond being graded is significantly larger or smaller than those of the master stone set, the colour cannot be graded as accurately. In grading larger diamonds the gemstone may show more colour and receive a lower colour grade, and with smaller diamonds they may grade better and receive a higher colour grade. As light passes through the larger diamonds, it has more material to go through and greater chance of absorption. In turn, the viewer sees that absorption as colour. With smaller stones the opposite is true. One can see the subjectivity in this already.

The subjectivity is furthered with the use of a five stone cubic zirconium master set because there is a colour grade that is skipped between each of the master stones. Here is what can happen. If someone uses a master set with the colours E-G-I-K-M and a diamond is found to have a colour just better than K but less than I, where does the gemmologist assign the colour grade? It could be said the colour falls in the middle and the stone could be given a J colour grade, but without an actual 'J' master stone, it is left up to the best judgment and perhaps more importantly the gemmologist's experience to assign the colour grade. If the stone is graded one colour grade higher than it actually is, at the wholesale level this could translate into several hundred dollars more than what the stone is actually worth. In looking at the difference in values for a 1.00 carat, J colour, SI2 clarity diamond versus a 1.00 carat, K colour, SI2 clarity diamond, both round brilliant cut, the J colour stone is worth about $400 U.S. wholesale more than the K colour stone. There are some definitive color grading instruments that could reduce the subjectivity of colour grading. While these are widely used by the major gemological laboratories, they are not widely used by jewellers.

One last thing about master stones is that the cubic zirconium varieties have been know to change colour over time. Usually the sets have a warranty for a period of five years to hold the colour accurately but ultimately this is another potential for the colour grade to be inaccurate. This is not to say that inaccurate colour grading occurs frequently through master stones that have discoloured over the years. Instead, the use of 5 piece colour master sets, the size difference in the diamonds being graded relative to the master stones, improper lighting, the difference in the way that each individual sees colour, and deliberate over-grading of diamonds are the greatest potentials for colour grades to be inaccurate.

Clarity

Unlike colour, where a gemmologist using a full master set can typically determine this grade, the criterion that is used to establish clarity grades is very much subject to the interpretation of the grader. In terms of values, the clearer the diamond or less flaws, the greater the value. First, internal and external flaws affect the clarity of a diamond. It is these flaws, the size, placement, colour, and number of these flaws that enables the gemmologist to establish a clarity grade. Flaws can be in the form of crystals inside the diamond, irregular growth structures, internal cracks, and several other forms. Sometimes the original surface of the rough diamond does not get polished away. This is referred to as a natural and this feature is found occasionally on finished diamonds. Typically external or surface flaws such as polishing marks affect the clarity very little and unless they are exceptional features they do not come into play except for grading flawless to internally flawless diamonds.

In actual fact, there are no truly flawless diamonds. Every diamond has some form of flaw or inclusion, however by International Diamond Council (IDC) standards, if these inclusions happen to be less than five microns in diameter under 10x magnification, for the purpose of clarity grading of diamonds, they are not considered to be a flaw[30]. As a general rule for diamond grading even if some sort of flaw (inclusion) is observed inside the diamond, if the inclusion happens to be less than five microns in diameter then it is not considered in the clarity grading and the diamond would then be considered flawless. This is not the case with every diamond grading school of thought but what the I.D.C. does is provide some specifics, something desperately missing in diamond clarity grading.

Having stated the ground rules for deciding what a flaw is, the criterion that is used for each particular clarity grade will be examined. For this were are going to use the G.I.A.' s clarity grades because it is the most widely used grading system in North America. The Gemological Institute of America and their gem grading certificates are well respected within the diamond and jewellery industry.

There are eleven clarity grades for diamond. These range from the best grade, Flawless (FL) to the poorest grade, Imperfect 3 (I3). Generally speaking, each clarity grade is defined by how difficult it is to observe inclusions within the stone. The poorest clarity grade of diamonds, Imperfect 3, has inclusions that are readily visible to the naked eye whereas with the top, flawless stones, inclusions would not be visible, even under 10x magnification. The fewer and smaller the flaws and inclusions, the more beautiful and rare is the diamond. As such a higher clarity value translates into higher per carat values. The general clarity grades that are assigned to diamonds using the G.I.A. grading standard are:

- FL and IF—Flawless and Internally Flawless (highest grades)
- VVS 1 & 2—Very Very Slightly Included 1 & 2
- VS 1 & 2—Very Slightly Included 1 & 2
- SI 1 & 2—Slightly Included 1 & 2
- I 1, 2, & 3—Imperfect 1, 2, & 3 (lowest grades)

The simplest way to remember these grades is to draw a line between SI (Slightly Included) clarity diamonds and I (Imperfect) clarity diamonds. Generally speaking, all diamonds above the line that are Slightly Included (SI) 2 grade and better have no visible inclusions when viewing the diamonds face up, through the crown[31]. Therefore these diamonds, SI clarity or better, require standard 10x magnification to see the flaws or inclusions. All the diamonds below the line in the Imperfect categories have inclusions that are visible to the naked eye through the crown. Taking this idea one step further and with all things remaining equal, if one was to look at an SI2 clarity diamond, face up, side-by-side with the finest quality diamonds of VVS or IF clarity, they would both appear the same to the naked eye. Again, this is because the flaws in the diamonds of SI2 clarity or better should not be visible to the naked eye. In looking at flaws that are so small and usually insignificant the clarity grades then becomes increasingly subjective.

There is a lot of ambiguity in the clarity description and this is where it becomes very difficult to establish a clarity grade that will withstand the scrutiny of other gemmologists. Only training and experience can lessen the ambiguity. Words and phrases within the grading criterion like 'observed with effort', 'noticeable', or 'obvious' are subject to each grader's interpretation. What is 'obvious' to one grade may be considered simply 'noticeable' to another. Each person sees the inclusions within a diamond differently and physically some have better vision than others. The type of inclusion, its colour and location in the diamond, how easily or quickly the inclusion is located, the size and number of inclusions located, and how it/they affects the diamond's brilliance and durability are all considerations for clarity. Add this, to establishing precise differences between each clarity grade and the ability to obtain a standard grade for a diamond's clarity is nearly impossible. Clarity grades are more like bandwidths that fade from one grade into another without a defining line between any of them. So in the end, clarity is the most subjective of the Four C's.

Grading diamond clarity is an opinion and one must look at the disclaimer found on every appraisal form that indicates that the grade of the diamond is simply the opinion of the diamond grader to understand that diamond grades, especially when it comes to clarity, are mere estimates. I once took a 0.50 carat diamond of I1 clarity, H colour, round brilliant cut and presented it to five different jewellers to see what grade they would give it. It received grades of I1, SI2 (x2), SI1, and VS2 from the different jewellers. The wholesale value of this diamond graded as an I1, H color, round brilliant is about $750 U.S.; as a VS2 its value is about $1,450 U.S.—nearly double. For several years I have delivered diamond workshops that educate people on these specific matters. In one part of the workshop, the candidates are given the basics of diamond grading, diamond grading criterion, and reference material for each of the clarity grades and are asked to grade clarity of several diamonds with a 10x loupe. Inevitably the range of grades that are returned are broad, often scaling between I2 to VS1 for diamonds that are actually I1 and SI2 clarity diamonds. The average person with no training or tutorials would find themselves with equal or greater variations of clarity on the same stones. I have made a point of illustrating this to clients by taking a picture of an SI clarity diamond and magnifying it 10x or more and then comparing it to other stones of various clarities equally magnified. Even side-by-side, it is difficult for the clients to decide what the clarity quality of the other stones are. On the positive side, the clients tend to go with an SI clarity diamond over the I clarity stones because there are no eye-visible inclusions. On the down side, the clients rarely want to go with stone clarity greater than SI1 because to the naked eye, it is not possible to tell a flawless stone from the SI clarity stone anyway. So it is understandable for them holding at the SI clarity stones, and other retailers encourage their clients to do this also. From a business point of view, it is great if a person can sell a higher quality stone. However, from the consumer's point of view, if beauty is what they are buying the stone for, the higher clarity means paying extra for something they can't even see. In speaking with a diamond retailer on this particular issue, he compared the buying of diamonds to buying a car, saying the SI clarity stones were like buying a base level Ford Crown Victoria, the VS clarity stones were like a fully loaded Crown Victoria, and the VVS clarity stones would be like driving a Lincoln Continental. My response to him was that with each one of those cars there is a noticeable, tangible upgrade, like power windows and other features such as leather seats, that comes with the higher cost. However, with the high clarity diamonds there is nothing tangible, no value added, no better features than what are received with the 'base model', SI clarity diamond. There is one thing added with the higher clarity diamonds and that is the higher cost of the stone. In the end, if a client chooses to spend more money on the stone, often jewellers will encourage them to put that money into a higher colour grade or a better cut, which is something tangible and noticeable.

It is interesting to look at diamonds of various clarity grades side by side to see the difference, or how little difference, there is between them. Below are four pictures of Princess Cut diamonds of similar carat weight, clarity and colour. These diamonds have been graded by reputable labs like G.I.A. and I.G.I (International Gemmological Institute). I say reputable because the grades handed down by these labs are among the most trusted in the industry. It is interesting to note that in real life these stones are approximately 5.5 x 5.5 millimeters in diameter which is about the size of an eraser end on a pencil. Despite how much they have been magnified for the photos, it is difficult to see any imperfections on any of the diamonds except the I1 clarity diamond. On the SI2 clarity diamond, the flaw in the bottom left corner appears as a dark cloudy spot. On the SI1 clarity and VS1 clarity diamonds even at this magnification it is difficult to see any flaws. Of course, in real life to the naked eye, one should not see the flaws in any of these stones except the I1 clarity diamond and even that will be difficult for the untrained person.

Princess Cut, 0.94 ct, E colour, I1 clarity
(flaws in top right and bottom left corner)

Princess Cut, 1.06 ct., G colour, SI2 clarity
(small flaw in bottom left corner)

Princess Cut, 1.01 ct., G colour, SI1 clarity
(difficult to find flaws, nothing apparent)

Princess Cut, 1.01 ct., G colour, VS1 clarity
(difficult to find flaws, nothing apparent)

For the SI1 and VS1 clarity diamonds, one would probably have to look at the stone from below with magnification and through the pavilion in order to see any flaws. Even under magnification there is nothing really apparent when looking down through the table of these stones.

Cut

The cut of a diamond makes reference to both the external shape of the gemstone but also the proportions that the diamond polisher has given the stone. In reference to cut, a person could say the diamond is a round brilliant cut diamond, making reference to the style or external shape of the diamond. Other shapes include triangle (trillion), square (princess), marquise, oval, heart, and emerald. Also, a person could refer to how well the diamond was cut in terms of its proportions. This is often also referred to as *make*. A diamond cut with excellent proportions shows optimum brilliance and dispersion. The pictures of the four (4) Princess Cut diamonds illustrate this well. Specifically, the SI1 clarity diamond has far better brilliance than its counterparts. It is important that the diamond is cut to specific proportions so that it reflects light perfectly. All the proportions of a diamond are measured relative to the diameter of the diamond (Figure 1, Diamond Proportions). If the crown is 54%, then that means the size of the crown facet is 54% of the diameter of the diamond. Similarly, if the diamond has a depth of 60%, then the total depth of the diamond (the distance from the table facet to the tip of the pavilion) is 60% of the diameter. The same holds true for the measure of the crown height, pavilion depth, and girdle thickness. They will be stated as a percentage of the diameter of the diamond. On a typical full cut diamond there are 57 facets with 33 facets on the top or crown of the diamond (Figure 2, Facet Names). This includes the table facet (1), kite or bezel facets (8), the star facets (8), and the upper girdle facets (16). The other 24 facets are on the bottom or pavilion of the diamond. This includes lower girdle facets (16) and pavilion facets (8). Sometimes there is an extra "58th" facet polished at the point of the pavilion, which is called the culet. The thin edge that is the separation point between the crown and the pavilion is the girdle. Occasionally the girdle is faceted but usually it is left unpolished and may even show small areas that are the natural surface of the original rough diamond. As previously mentioned, when parts of the original rough diamond surface are left unpolished on the finished diamond, these parts are aptly referred to as "naturals".

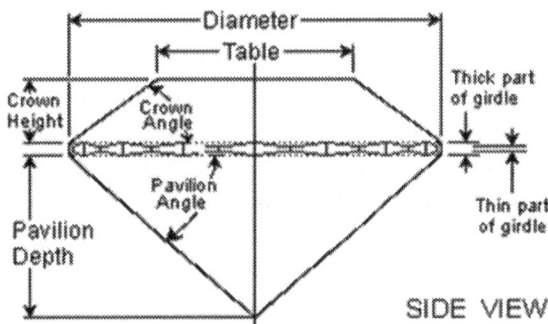

Figure 1: Diamond Proportions

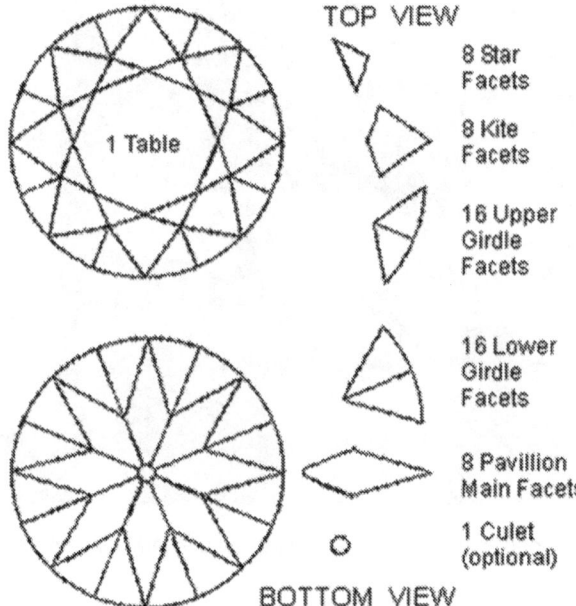

Figure 2: Facet Names

Source: Wikimedia[32]

As cut proportions deviate from what is considered an excellent cut, then they are assigned lesser-cut values. Again, the more beautiful the diamond, the higher the value. Excellent cut diamonds are more rare than good cut diamonds, as cutters tend to cut most diamonds for good cut proportions. This is because the good cut diamond's proportions result in greater weight retention from the rough diamond than an excellent cut diamond's proportions. This is often the diamond cutter's dilemma; to cut for perfect proportions or for greater recovery weight from the rough diamond. As such, excellent cut proportions, also known as *ideal cut* diamonds, command higher per carat values. A poor cut diamond translates into a less beautiful diamond and therefore lower per carat values. In addition, because of irregular proportions, a diamond of lower carat weight may actually have a larger diameter than a diamond of a higher weight. I have seen several diamonds in the 0.94–0.99 point weight class that are larger in diameter than some diamonds over 1.00 ct.

Common finished diamond shapes of round, radiant, oval, emerald, and pear shapes. Photo courtesy of GIA.[33]

An experienced diamond grader can assess and establish a cut grade of a finished diamond immediately upon viewing the diamond. This is a result of seeing the optical properties of thousands of diamonds. A well-cut diamond displays the ultimate of brilliance and fire and seems to have a life of its own. On the other hand, a poorly cut diamond has correspondingly poor brilliance and fire and these stones are often referred to as being 'lifeless', having dead spots, or a 'fish eye' if the girdle is too thick. A mathematician named Tolkowsky established the best-cut proportions that maximized the brilliance and fire of a round diamond in 1919. His proportions are all based on the girdle diameter of the diamond. For instance he established that the height of the crown should be 16.2% of the girdle diameter, the depth of the pavilion should be 43.1% of the girdle diameter, and the diameter of the table should be 53% of the girdle diameter, and total depth of approximately 60% of the girdle diameter[34]. There are other types of brilliant cut diamonds including the Ideal Brilliant, the Practical Fine Cut, Scandinavian Cut, and the Parker Brilliant that all have different proportions than the Tolkowsky brilliant[35]. This makes judging cut even harder because the proportions that may be excellent for one cut style may not be excellent for another cut style. One style may have a greater return of light or brilliance, while another style may have more fire. Here the beauty is definitely in the eye of the beholder. Typically diamonds are cut to proportions with slight deviations from those specified by Tolkowsky. The deviations in proportions or tolerances allow the diamond grader to assign a grade to the cut of the diamonds based on the amount of deviation from the Tolkowksy proportion. For instance, the tolerances for a diamond that is considered a good cut could have proportions such as a crown height of 14%, pavilion depth of 47%, and table diameter 64%. Diamonds in this tolerance range are quite common and are known in the industry as a 60/60 cut, meaning it has approximately a 60% table width and 60% total depth. These 60/60 diamonds sacrifice very little in terms of brilliance and fire from a diamond cut to the specific Tolkowsky proportions, however, the increased table diameter results in a higher weight retention than one cut with Tolkowsky proportions. The result could be a diamond that weights 1.01 carat but if it had been cut with Tolkowsky proportions, it may have only weighed 0.98 carat or less. As known from the non-linear scale for diamond values, this could result in several hundred dollars per carat more for the diamond that eclipses the 1 carat threshold.

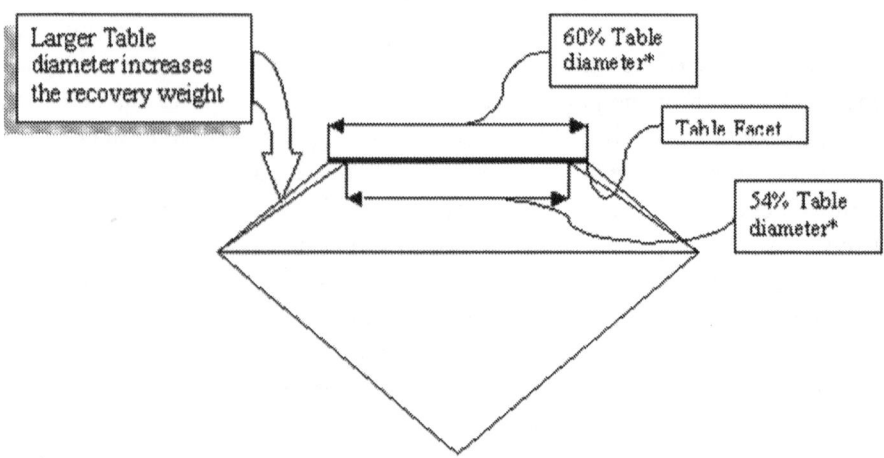

Larger Table diameter increases the recovery weight

60% Table diameter*

Table Facet

54% Table diameter*

*measurements are not to scale

These submissions get people wondering about diamond's they've purchased or wondering how to nail down exactly what is being bought when purchasing a diamond. The reasons criminals like this commodity are already emerging.

Having discussed the valuators that are used to qualify the characteristics of a diamond, it is now possible to define a price for any diamond. Just how valuable can some of these diamonds get? There are very few top quality one carat finished diamonds available for sale in any one year and relatively speaking, any 1 carat gem quality diamond is rare. Comparing values of 1 carat diamonds using our table, at the very highest quality we have a 1-carat, D colour (the best), FL—flawless clarity (the best), diamond with good cut proportions sells for approximately $17,000 U.S. At the lower end a 1 carat, M colour, I2—Imperfect 2 clarity, with good cut proportions sells for $1700. U.S. One ounce of low end diamonds could easily price out at approximately $300,000. Canadian wholesale and one ounce of the highest quality 1 carat diamonds would be worth about 3.5 million wholesale and would fit into your pocket. The reality is it would be very difficult to put together 155 x 1.00 carat diamonds at the flawless, D—color range, but stones could be collected in high colour/clarity parameters.

So how does the value of diamond compare to gold? First, speaking about diamonds and carat, it refers to weight, but when talking about karat it refers to the purity of gold. 24 karat gold it is pure gold, 12 karat gold is only 50% gold, 18 karat gold is only 75% gold, and the rest is some other metals that have been mixed with it. So, it is possible to have 1 carat weight (0.2 grams) of diamond and possible to have 1 carat weight (0.2 grams) of gold. On the other hand it is possible to have 24 karat gold (pure gold), but one cannot have any kind of karat diamonds. To move along, one troy ounce of gold is worth approximately $550 U.S or $675 Canadian. We divide this by 155 carats, which is equal to one troy ounce in weight, which means that 1-carat weight of pure gold (24 karat) is worth about $4 Canadian.

1 carat top quality diamond approx. $22,500 (Can.) vs. 1 carat pure gold (24 karat) $4 (Can.)

Previously, fancy coloured diamonds were discussed. These are diamonds that show vivid colours of yellow, green, blue, red and pink. As a matter of fact, diamonds can come in nearly every colour. These diamonds are so rare that they command prices typically beyond that of high-end white (colourless) diamonds. In recent years one fancy red diamond of approximately one carat sold for over $1 million U.S. These diamonds have particular colouring agents or crystal structure irregularities that give them these distinct colours. There are treatments available to artificially colour diamonds to a fancy colour, and these artificially coloured diamonds are now common. If you see a fancy coloured diamond, don't believe it's a genuine untreated fancy colour diamond unless it comes with an independent certificate from a reputable lab; then call the lab to make sure the certificate isn't a forgery. By the same token, if you find a D colour, Flawless diamond, be careful with these as well. Unless they come with a certificate from a reputable lab, like G.I.A, I'd recommend staying away from them.

Various fancy coloured diamonds (blue, yellow, green, and pink) and in both rough and finished form. Photo courtesy of GIA[36]

4

Laws and Industry Controls

In Canada, the production, cutting, wholesaling, and possession of diamonds goes on largely uncontrolled. The argument for and against greater controls on diamond is strong for both sides with the balance internationally starting to tilt toward greater controls. This is seen in new legislation designed to track the sales of diamonds and jewellery. This is similar to legislation that is already in place that allows governments and law enforcement to keep track of individual's large financial transactions. It is also apparent in the international agreement known as the Kimberley Process, which was created to thwart the trade and sale of conflict or blood diamonds. As a commodity like any other product, diamonds are produced and retailed, used to machine tools, used in medical, electrical, and optical equipment, and provide jobs, prosperity and enjoyment.

On the other hand, diamonds can be corruptive. They are used as a currency, are valuable to criminals and terrorists, fuel civil conflicts, and are a primary target of criminal acquisitions. The Kimberley Process and associated international measures, and corresponding legislations to curb the sale of conflict diamonds are supporting greater controls on diamonds. In the United States as of January 1, 2006, diamond dealers are required to report diamond transactions exceeding $10,000 cash in a measure to track the potential movement of diamonds being used as currency and in money laundering schemes[37]. Other countries like Canada are working towards implementing control measures on diamond transactions similar to those that the United States has recently enacted. This provides more evidence that globally, governments are moving to establish tighter controls over diamonds. At the national level, several diamond producing countries have laws and regulations that control the production, possession, distribution, and cutting of diamond. In Canada, the only diamond specific legislation is the Export and Import of Rough Diamonds Act (EIRDA). This legislation is not unique to Canada and is the result of international legal coordination that came from the United Nations Kimberley Process. The Kimberley Process is an international accord designed to stop the proliferation of conflict diamonds. In Canada, aside from EIRDA, there are only a few pieces of Canadian legislation that indirectly deal with diamonds.

The Criminal Code

Under the Criminal Code there are few obvious sections that come into play with crimes committed in relation to diamonds, gemstones, and jewellery. Both theft and fraud are the most common examples in relation to basic theft of jewellery or fraud in relation to selling a fake diamond as a genuine stone. Perhaps even Passing Off Section 408 of the Criminal Code could be used if someone tried to sell a non-Canadian diamond as a genuine Canadian diamond. Robbery is of course considered in cases of theft with violence. However, there is one Section that specifically targets crime related to rough diamonds and any other valuable mineral. Under Section 394.1 of the Criminal Code, there are provisions to charge a person for criminal possession of a valuable mineral or fraud in relation to a valuable mineral. The valuable mineral in this case must be a mineral in its raw unaltered form and have a value of $100 per kilogram or roughly $45 per pound. At $45 per pound this would even include industrial quality diamonds. The criminal code is administered and enforced by all police forces in Canada, yet there have been few such cases reported.

Customs Act

The Customs Act controls diamonds and jewellery in the same manner that it controls any other commodity in that it requires citizens to report out of country purchases and any imports of these purchases into Canada. This is also the usual place where the tax is collected because once people report the out of country purchase, they are required to pay taxes on gemstones and jewellery. Generally speaking the taxes on diamond are the Goods and Services Tax (GST) and Provincial Sales Tax (PST) for non-commercial imports. Until recently there was also an excise tax on diamonds that amounted to 6% of the value for duty. The Canadian Border Service Agency (C.B.S.A.) examines goods coming into Canada and collects any taxes including the GST and PST on behalf of the federal and provincial governments. The C.B.S.A. is the first line of defense in enforcing these Federal Acts and with this they have an incredibly difficult job. Imagine flying into Vancouver or Toronto International Airport with 300 other people on the international flight and now it is the time to clear customs. Add two or three other planes that have just arrived and there are hundreds if not thousands of people needing to clear customs. In addition, the C.B.S.A. officers have to process all these people very quickly as other planes are now coming in with many connecting immediately to other domestic flights. There may only be five or six customs agents on duty to check all of those people clearing customs and they do this based on little more than a customs declaration and a person's word. It is amazing that with such little information provided to these officers and the time constraints that they are able to uncover any contraband or smuggling at all. These officers are trained to identify those people that exhibit suspicious signs, which leads them to have a closer look. They do an excellent job identifying those carrying smuggled goods, counterfeit, or prohibited products. Within any calendar year there are newspaper accounts of Customs Officers across the country whether Vancouver, Halifax, Toronto, or Calgary, making a seizure of jewellery from a smuggler. Recently, a Canadian jeweller was charged with smuggling over a quarter million dollars in U.S. currency along with over 3,000 pieces of jewellery and steroids out of the United States[38]. With diamonds, gemstones, and jewellery, easily concealed items that are highly taxed, this is becoming an even more difficult task. Compounding the issue, law enforcement officers typically do not have any formal training on these commodities and there is certainly no mention of special units, or jewellery or gemstone targeting on the Canadian Border Services Agency web site. Yet, regular seizures have been made of people smuggling in diamonds, gemstones or jewellery. Items like high karat gold is smuggled into Canada in an effort to avoid high taxes. Other items like gemstones and pearls may carry a 13%–21% total tax. With the high tax comes the added incentive to smuggle to avoid the tax[39].

Export and Import of Rough Diamonds Act

This Act was created as a measure to stop the sale and trade of conflict or blood diamonds. In brief, these rough diamonds are known to come from countries in West Africa that are or have been experiencing civil conflict. The diamonds themselves or proceeds from their sale are used to purchase arms and munitions to support the conflict. These civil conflicts result in death, severe casualties, and other gruesome atrocities to civilians. The name blood diamonds was minted on the flip side to conflict diamonds in making reference to the blood shed. The United Nations proctored an action to reduce the sale of conflict diamonds in what is known as the Kimberley Process (KP). This action is largely a result of lobbying by non-government organizations such as Global Witness who has led the charge in bringing this issue to the attention of the world. Over 50 nations around the world have signed onto the Kimberley Process and many have enacted legislation in their respective countries to implement the Kimberley Process protocols. In short the Kimberley Process protocols prohibit the sales of diamonds that are not government certified. This government certification of the rough diamonds is basically a certificate of origin and warranty that the stones are not conflict diamonds. The sale or trade of rough diamonds is not permitted to countries that have not signed on to the Kimberley process.

In Canada, the Export and Import of Rough Diamonds Act, which came into effect January 1st, 2003 was enacted to implement the Kimberley Process protocols. In short, a rough diamond producer in Canada and wanting to export rough diamonds must obtain a Kimberley Process certificate from Natural Resources Canada to do so. On the other hand and aside from limited exceptions, a rough diamond importer will require a Kimberley Process Certificate from the country of origin in order for it to pass through Customs. Even if the rough diamonds are properly declared, if they are not accompanied by a Kimberley Process certificate, the importer could be charged with an E.I.R.D.A offence and lose the diamonds.

The first ever charge in Canada under this Act came in Toronto, Ontario when an astute CBSA officer was checking individuals deplaning a flight from Africa. The criminal was sent to secondary, where the CBSA officer is able to thoroughly examine the personal belongings of a suspicious person, and it was here that he discovered nearly $20,000 worth of rough diamonds on his person. The smuggler was charged with importing diamonds contrary to EIRDA, convicted in court, and received a small fine and forfeiture of the diamonds.

In normal instances, anything forfeited to the crown becomes crown assets, which are then sold later at government auctions, provided it is nothing illegal. Of course it would not be in the public's best interest or that of the Kimberley Process to sell the conflict diamonds back to the public. As such, within the Export and Import of Rough Diamonds Act there are provisions for diamonds that are seized and forfeited through the courts or otherwise through the Act that allow the rough diamonds to be used for specific purposes such as scientific study.

This provision is actually very important because the RCMP, in their forensic lab in Ottawa, have been working on what is known as laser ablation mass spectrometry. It is known to all those who are not scientists as diamond profiling. The idea is that diamonds from each particular mine in the world have a chemical signature or a fingerprint. Diamond is made of pure carbon, but within the diamond at levels measured in parts per trillion, are trace elements. It is believed that these trace elements are like fingerprints and are tell tale of not only what country a diamond comes from, but what specific mine site. The potential benefit from this research is enormous. For one thing, being able to say precisely where a diamond comes from would help curb the international sales of conflict diamonds. Another benefit particularly for Canadian mined diamonds is being able to say definitively that a diamond sold to a consumer is in fact Canadian. Despite the labeling of some diamonds as Canadian diamonds, the process that certifies diamonds as being Canadian is not flawless or absolute and this technology could help. Last, there is a potential benefit for Canadian law enforcement. Prosecutions under the Customs Act for smuggling of diamonds could be much simpler if the prosecution could declare that the diamonds definitely were not mined in Canada. In this instance the diamonds could be rough or finished because the test that is done on the diamonds is so unobtrusive that it would not even be noticeable unless the diamond was under high magnification.

In terms of the proliferation of conflict diamonds, one of the big questions for law enforcement is how easy is it to get hold of rough diamonds and how easy is it to get them into Canada from another country? Unwittingly I may have answered this question a few months after the Kimberley Process was initiated and both Canadian and U.S. legislation had passed. I contacted a couple of rough diamond dealers in the United States to see how easy it would be to purchase some rough diamonds and get them to Canada. With the new legislation in place in Canada and the United States, I sent a message out to both companies seeking to purchase rough diamonds. The text of the message and responses to them appear verbatim below:

Hello

I'm interested in purchasing about 10 sawables now. Maybe more later. As well as some of the cubes you have.

There was a new law passed in Canada and all rough diamonds that come into Canada must come with a certificate of origin.

Do your diamonds come with certificates?

Thanks,

Kelly'

As indicated in Chapter 2, sawables describes types of rough diamonds. Sawable describes a rough diamond that can usually be cut into two pieces. Then each piece can be cut and polished into a finished diamond. These are nice stones to work with. Cubes, are more risky and there is no guarantee that a finished stone or any finished diamond will be obtained from the rough material. I threw these terms in with hopes that the seller would respond to my message and feel comfortable with whom they were dealing. Many diamond dealers, especially rough diamond dealers, don't want inquiries from tire kickers or gem enthusiasts. They want to deal with serious buyers only. At the same time I didn't mention what carat weight of diamonds I was seeking or even how many cubes I was looking for, because this could have raised some flags for the dealer. My sense of this is that they are rather small time rough dealers, perhaps even criminal, and they just don't care. Also, I specifically asked the dealers if the rough diamonds would come with a certificate of origin. I received a response from both sellers and I was astonished at what they said and what I learned from a simple inquiry.

Here is the response I received from one of the rough diamond dealers;

'Good Morning,

I express the diamonds in (to Canada) under the gift clause of a rough mineral through FedEx.

I do this for insurance reasons and to get it through Canadian Customs quickly.... or they sit on it for weeks!

I enclose no paperwork with the item, it is sent as a letter post so the item can not be associated with paper or price.

I know the shipping seems high but believe me, you will not get taxed and it will get to you in 72 hrs.

I have many Canadian clients and this works without fail!

Cheers....'

(seller's name omitted)

This was an eyebrow raiser. First, there was no response to the question about the certificates so it is assumed they don't supply a Kimberley Process certificates with their rough diamonds. Of course this is an offence under the Export and

Import of Rough Diamond Act in Canada and against the law in the United States. Secondly, the contents of the package are misdescribed as a gift rather than a personal or commercial import of goods. This is a potential offence under the Customs Act. No paper or price is done specifically so the CBSA has a difficult time trying to put a value for duty on the diamonds and this way they slip it through the system quickly and avoid taxes. Last, this vendor claims that this is done regularly and without fail. It begs the question of how much is going through the border? As a point of interest, I checked recently and this dealer is still selling rough diamonds. The second response, reprinted below, was even more surprising.

NO

WE SHIP AS EXOTIC FISH TANK STONES OR IN MANY CASES EXPERIMENTAL SPECIMENS. WE SHIP TO CANADA 3X'S A WEEK.

THANKS

(seller's name omitted)

YOU MIGHT BE TALKING ABOUT CUT DIAMONDS, I'VE NEVER HAD THIS QUESTION BEFORE

In this case the rough diamond dealer explains that he purposely misdescribes the diamonds as exotic fish tank stones or experimental specimens. Look at a bag of rough diamonds below, it is plausible that this explanation could fool someone. Like the first vendor, he ships to Canada on a regular basis, does not provide Kimberley Process certificates for his diamonds, and apparently hadn't even heard of the Kimberly Process or requirement to supply certificates with his rough diamonds. Generally speaking, with two small time rough diamond vendors as a reference, there seems to be a thematic disregard for the export/import laws in several respects as well as the Kimberley Process.

Bag of Rough Diamond a.k.a Fish Tank Gravel?

Ultimately I did not make a rough diamond purchase from either of these vendors and I took my quest to purchase rough diamonds to Antwerp, Belgium. I contacted a rough diamond dealer operating in the heart of the diamond dealing district of Antwerp. This dealer operates out of one of the diamond bourse buildings and the security is very tight. You expect this when there are hundreds of millions of dollars worth of rough and finished diamonds in the building. There are guards with guns, close circuit cameras in buildings and outside on the street, protective glass and the likes. Even the public trash bins are bomb proof to prevent anyone from using them as a place to hide and then detonate a bomb. You don't just walk into these places you make appointments and you have to drop your passport off before security will let you into the building. In short I made a deal on a small quantity of rough diamonds, shook hands and when I asked the dealer to arrange for Kimberley Process certificates the deal went sour. In the case of small purchases he was not prepared to make these arrangement and told me to simply smuggle the diamonds out of Belgium and into my country, concluding with "this is what all my small buyers do". While I understand that there may be an added cost as a function of the work required to facilitate the exporting of the stones on my behalf, what I don't understand is why this diamond dealer would be advocating me to break both Belgium and Canadian law.

Equally disturbing is how these diamonds, like other products including contraband substances, simply slide through our boarder. To be fair, there is the possibility that the industry as a whole may be unaware of new diamond specific legislation, or of a misunderstanding of the laws that apply to diamonds. This doesn't preclude someone from not following the law as knowing the laws that pertain to a particular industry is fundamental to running a business in that industry. Each particular case is different. As shown in the next paragraph, application of the laws concerning diamond can be difficult to interpret, even for those who are supposed to enforce them.

Cultural Property Export and Import Act

This Act is very broad and has been around since the 1970's, long before diamonds were mined in Canada. It is designed to protect items from Canada of significant aesthetic, historic, scientific, and cultural value, including works of art by Canadian artists, Indian Head Dresses, Dinosaur bones, Military artifacts, and more. There are eight protected groups under this Act and Group 1 is 'articles removed from the soils and waters of Canada'. Within this is found the minerals definition that in short, includes all valuable minerals whether rough or cut and polished. The Cultural Property Export and Import Act (C.P.E.I.A.) specifically defines a mineral as "an element or chemical compound that occurs naturally in the soil or water and includes crystals and naturally occurring metals, and gemstones whether or not polished or faceted by a person or persons. It does not include minerals, ores and concentrates intended for industrial use, or a carving or sculpture made by man from minerals". This obviously includes Canadian diamonds or any other gemstone found in Canada. As such, diamond, sapphire, emerald, and any other mineral that meets the specific parameters and value thresholds of the Act may require an export permit. The threshold for minerals protected by the Act and requiring a permit to be exported, is;

a. a single mineral specimen of a fair market value in Canada of over $2,000,

b. a collection of 10 or more mineral specimens of a fair market value in Canada of over $5000 from the same source (mine), quarry or locality,

c. mineral specimens in bulk, recovered from a specific mineral occurrence, weighing 225 kg or more of any value, and

d. meteorites and tektites of any value

Up until 1998, this Act had little to do with valuable gemstones in Canada aside from ammolite or nephrite (aka British Columbia jade). However, with diamond also falling under this Act, the scope and amount of minerals that require export permits has risen dramatically. As mentioned before the C.P.E.I.A. controls all valuable minerals not just ammolite and this includes diamond. Therefore, depending of the interpretation, all Canadian diamonds rough or finished that meet the individual threshold values of $2000 or the group threshold value of $5000 are subject to export control under this Act. These are only diamonds that are not mounted in jewellery. This means that there are tens of thousands of Canadian diamonds per year that fall within the threshold parameters of this Act. This is a huge amount of diamonds that require Cultural Property export permits. However, to streamline this process a company that exports a lot of Canadian diamonds can simply apply for a general permit to cover all their diamond exports for the year. In speaking with all the people I know that export Canadian diamonds, there is zero compliance with this Act. People do not know about this Act as it pertains to Canadian diamonds or their duty to comply with it. There are new deposits of other gemstones that are under development in Canada including emerald and sapphire. If and when these gemstone industries begin to export their products and if they meet the threshold requirement, they will also have to comply with the CPEIA. Although this Act specifies that minerals are to be controlled as Cultural Property, the spirit of this Act is to control those items that have a significant historical, scientific, aesthetic, or cultural value to Canada. Depending on one's perspective, the average diamond or gemstone may not be considered a significant value aesthetically or scientifically. By the same token, the spectrum of what some people consider aesthetically pleasing or scientifically important is wide open and subject to interpretation.

The Copyright Act

Fake brand name jewellery, fake Canadian brands of diamonds, fake gemological certificates, and fake brand name jewellery packaging are just a sampling of the international copyright problems. All levels of criminals, including terrorists, are involved in copyright crimes through producing and selling fake, brand name merchandise[40]. Brand names and trademarks are property owned by companies and allow the companies to market their product as something different than other companies producing similar items. Generally speaking, the only difference between Canadian diamonds and diamonds from anywhere else in the world is the branding of the diamond as being Canadian. Aside from impurities measured at the atomic level, diamond is diamond, whether Canadian, Russian, South African, or whatever the country of origin. What some Canadian companies do to separate their diamonds from those of other countries is to brand them as 'Canadian'. The way they brand them is to laser etch their unique trademark on the girdle of the diamond. Through this marketing approach these companies are able to separate their products from the competition with a great deal of success. In fact, the Canadian branded diamonds have been marketed as the Fifth C. This suggests that along with the four universal diamond valuators that there is a fifth valuator added to the diamond by the fact that is has been mined in Canada. Criminals wanting to capitalize on the success of Canadian diamond branding could have a similar trademark likeness laser etched onto their diamond's girdle for about $30. If such cases occur outside of Canada they are difficult for Canadian authorities to investigate. Other criminals have exploited the trusted gemological grading certificates of the Gemological Institute of America by producing counterfeit certificates for the gemstones they are selling[41]. Gemmological certificates usually come with large diamond purchases and coloured gemstones. However, for the most valuable diamonds and precious gems, certificates from the G.I.A. and other trusted laboratories are what is use to ensure that the purchase is authentic and meets the quality that the seller specifies. Frauds in this regard have occurred with fake certificates, falsified to show a higher than actual quality of the gemstone. Similarly, Tiffany & Co. found counterfeit Tiffany jewellery being sold on the Internet and had to get a court injunction to stop the seller from making further sales of the counterfeit products. It is estimated that in this instance alone the seller generated revenues in excess of $500,000[42].

The Competition Act

The Competition Act is administered by the Competition Bureau of Canada with the sole focus of the Act to ensure fair competition in the market place. Under this Act there are provisions for charging a person for engaging in deceptive marketing practices. This includes a spectrum of activities including someone who holds a perpetual 50% off sale, to someone who makes false or misleading representations. In terms of diamonds, the possibilities for criminals to mislead the public in respect to quality is enormous. However, the Competition Act has no sections that deal precisely with diamonds, gemstones, or jewellery. The Competition Act, like the Customs Act, deals with these commodities as it would the sale and marketing of cars or televisions. However, for the diamond and jewellery industry the Competition Act can make reference to guidelines that set out acceptable standards and business practices for the industry. These guidelines have been developed by the industry in conjunction with government agencies in an effort to prevent deception and misleading opportunities. They are used by the Competition Bureau to deal with diamond and jewellery issues that arise, but on their own are not legally binding. While the Competition Act is a government statute with lawful authority, the guidelines (covered next) are not, and are in fact just that—guidelines.

"The Guidelines"

In the absence of legislation specific to diamonds, coloured gemstones and pearls, the diligent efforts of the diamond and jewellery industry has created the Canadian Guidelines with Respect to the Sale and Marketing of Diamonds, Coloured Gemstones, and Pearls. For simplicity sake it will be referred to as the 'Guidelines' from here on in. The Guidelines were created by the "Jewellers Vigilance of Canada Inc. special committee in consultation with industry members and the Competition Bureau and with reference to other internationally recognized standards"[43]. This alone, speaks volumes for the integrity of the Canadian diamond and jewellery industry. These Guidelines, designed to bring an industry standard to this segment of the diamond market, are very specific but like the title suggests, they are guidelines, with no real legal bite to it. It also means that whatever is written as law takes precedence over the Guidelines. The scope and application of this guide is set out at the beginning of the guide and literally spells this out for the reader. Within the scope it details that this is but a guideline and developed in consideration of the Competition Act. It further reads that the guidelines are for assistance purposes only and are not binding. It goes on further to advise the reader that they should request an opinion from the Competition Bureau (for a fee) to determine if their proposed business practice or plan would violate any portion of the Competition Act. This is problematic. If the Competition Act alone does not set out its own law or code specific to the marketing of diamonds that is binding, then really there is little that is enforceable. Here lies a huge opportunity for criminals to conduct business in any manner they chose relative to the marketing of diamonds to consumers. Basically what we have in Canada is a completely unregulated diamond market. Within this ambiguity lies the perfect opportunity for criminal exploitation of diamonds, as they will exploit any regulatory weakness or loopholes. Most of those in the industry that I have spoken with know about the Guidelines and say they follow them, however, I have also seen deceptive marketing of diamonds that clearly violates the guidelines.

The following Scope and Application are excerpts directly from The Canadian Guidelines with Respect to the Sale and Marketing of Diamonds, Coloured Gemstones and Pearls page 1, revised edition 2003.

SCOPE

The definitions and misuses of terminology outlined in these guidelines were developed in consideration of the *Competition Act* (a portion of which can be found in *APPENDIX 1* of these guidelines) that contains prohibitions against false and misleading representations. Adherence to the nomenclature contained in this document will assist jewellery industry members in their obligation to ensure compliance with the legislation and to provide consistent and meaningful information to consumers.

The guidelines are for assistance only and should not be considered as binding on the Commissioner of Competition. All methods of making representations, including printed or broadcast advertisements, written or oral representations, audio-visual promotions, Internet and illustrations are within the general scope of these guidelines.

The examples contained in these guidelines are for the purpose of illustration only and are not intended to provide an exhaustive list of acceptable or prohibited practices. The Competition Bureau facilitates compliance with the law by providing legally binding written opinions subject to fees. Advertisers are encouraged to request an opinion on whether the implementation of a proposed business plan or practice would raise an issue under the *Competition Act*. A specific opinion will be based on information provided by the requestor and will take into account previous case law, prior opinions and the stated policies of the Bureau.

Finally, readers should note that the misleading representations and deceptive marketing practices provisions of the *Competition Act* comprise only a portion of the relevant law in Canada. Most provinces and other federal departments and agencies also administer legislation dealing with advertising and marketing practices. These guidelines do not to provide information on this other legislation.

APPLICATION

In general, these guidelines apply to anyone promoting, directly or indirectly, the supply, use, description, identification, sale of or trading in any gem, carving, jewel, item of jewellery or work of art containing diamond, gemstone, pearl and related materials.

If a criminal enters the jewellery industry to sell diamonds in a manner that is contrary to the Guidelines, there is little to stop him from doing so. It is usually a legitimate jeweller that notices unscrupulous practices or a consumer that has been swindled that finally tips off the Competition Bureau. When the Competition Bureau checks on the criminal to see what he is doing wrong, they have to decide if what he is doing wrong is, in fact, an offence under the Competition Act. If it is, they may issue a notice to the criminal vendor outlining what it is that he is doing wrong and command him to cease the offending business practice. If he then continues his offending business practice, it is possible he could be charge under the Competition Act. Because the Guidelines don't correspond directly to any infraction under the Competition Act, it can be difficult for a Competition Act charge to stick. I have often noticed in jewellery store windows, a diamond ring listed for sale with the appraisal certificate. The appraisal certificate will say for example, that the diamond ring is worth $10,000 but the ring is actually retailing for $7,500. In these instances, the appraisals are being used as a tool to sell the ring by implying to the consumer that they are getting a better deal. Based on the guidelines this is an unacceptable marketing practice. The Guidelines Section D20 says "**It is contrary to the purpose of these guidelines to use an appraisal value as a selling tool.**" The same situation occurs if a clerk, selling the same $7,500 ring, states to the customer that it appraises at $10,000. Yet, I am unaware of someone in the industry being criminally charged under the Competition Act for this type of deceptive practice. In reality, the Guidelines need to be enacted into law if it is to have any effect on unscrupulous marketing practices and to garner public confidence. The same deceptive marketing practices can be found in the United States, however, they have enacted legislation that specifically deals with deceptive jewellery industry practices. These laws are found under the federal trade regulations. While these laws have not eliminated deceptive jewellery industry practices in the United States, they do regulate the industry, and provide law enforcement and the public with a tool to engage the criminals.

"The Code"

The Voluntary Code of Conduct for the Authentication of Canadian Diamond Claims is another voluntary compliance standard for the industry. It will be referred to as the 'Code'. Basically this Code is for anyone who is selling diamonds and wants to be able to call them or market them as Canadian. Of course this means that the diamonds have to be mined in Canada in the first place. This Code outlines the steps that vendors must take to be able to say that the Canadian diamonds they are selling are in fact Canadian. The main focus of the Code is the tracking of diamonds from mine to market. This is "based on a paper trail and a chain of warranties" through producer and manufacturers so the diamonds can be tracked from consumer to mine site[44]. The last step of the process is a laser etched serial number on the girdle of the diamond. This serial number is the tracking number for the diamond. The Code evolved from the collaborative efforts of a broad cross-section of the industry including the diamond mining sector, cutters and polishers, retailers, the Canadian Jewellers Association and Jewellers Vigilance Canada as well as the RCMP and other government stakeholders. The Code is administered by the Canadian Diamond Code Committee, a non-profit volunteer industry body comprised of representatives from the mining, cutting and polishing, wholesale and retail sectors, Jewellers Vigilance Canada and Canadian Jewellers Association as well as a national consumer association. This Committee is responsible for the maintenance of the Code Signatories' Registry and the 1-800 hotline service. The Committee responds to authentication requests from consumers by obtaining or confirming required information under the Code to track the diamond from the retailer to the mining company. If a consumer or interested person has a question about the code or wishes to track the authenticity of his/her diamond, this volunteer committee will do so for a small fee. If someone wants to sell Canadian diamonds they become a signatory to the Code, which means they agree to abide by its standards. The primary penalty for conducting business contrary to the Code is being removed as a signatory to the Code.

> "A Code signatory found in non-compliance with this Code will be removed as a Code signatory from the Code Registry. Failure to authenticate diamonds represented as Canadian may lead to an investigation by the Competition Bureau and subsequent enforcement action pursued under the false or misleading representations provisions of the Competition Act."[45]

5

The Appeal to Criminals

When the Great Blue was stolen from French Royalty, the motivation was pure and simple—profit. It was not about owning the finest diamond in the world but about how much the diamond would fetch. Surely those who perpetrated the crime were not the ones who in the end had the Great Blue in their vault. That is they didn't keep the beautiful blue diamond simply as a keepsake. In 2000, criminals tried to steal a hundred millions dollars worth of diamonds on display at the millennium dome in London, England, but expected only to reap a few hundred thousand dollars from the buyers[46]. Similarly, in February 2003, organized criminals cracked the central vault depository used by the Antwerp diamond merchants and approximately $150 million dollars worth of diamonds and diamond jewellery was quickly stolen. The criminals were apprehended within a few days of the heist, yet the loot was nowhere to be found. The underlying theme here is that stealing diamonds has been very profitable for criminals for centuries. It's not about owning the diamonds; it's about how much money they can fetch.

In today's world the criminal use of diamonds is much more varied and complex than a simple theft and conversion to cash via sale. Diamonds are now used to perpetrate frauds, generate criminal financing, money laundering, storage and movement of proceeds of crime, or traded directly for contraband. A U.S. government study of several commodities including illicit drugs, cigarettes, weapons, gold and diamonds, found that only diamonds have the ability to earn money, move money, and store money.[47] As such, the appeal of diamonds to criminals is greater now than ever. Historically, diamonds and jewellery have been items highly sought after by criminals. However, some recent statistics out of the United States suggest that this behavior is changing and the trend is for increased criminal desire for jewellery and precious metals. U.S. statistics show that from 1999 to 2003, the value of jewellery and precious metals stolen by criminals moved from 4th place to 2nd place among other stolen commodities like stereo equipment, cash, electronics, and cars[48]. By value, only automobiles beat this category. There is no reason to suggest this would not be the same for Canada. The value of the stolen jewellery and precious metals has risen approximately 45% in those same five (5) years from $709 million to just over $1 billion[49]. This makes jewellery and precious metals the fastest growing of all the commodity categories. There is no side stepping the fact that criminals love diamonds and jewellery.

The beautiful thing about statistics and quantitative analysis is that they provide current status and historical data. Although statistics or quantitative analysis is able to indicate what is happening, it doesn't provide information as to why it is happening. Today, in the Intelligence Lead police environment, investigators can far surpass what the statistics provide and tap into non-statistical qualitative information sources. Where the quantitative information can uncover what is happening, the qualitative information can illuminate why things are happening. When the two are analysed together with a little creative thinking and prediction, investigators are provided with a greater awareness of what is going on in the criminal world. Perhaps even understand how things are happening.

In its simplest form, this is how Criminal Intelligence (C.I.) is produced. Through C.I., law enforcement is able to more accurately target criminal activity and facilitate a more forward looking, proactive, effective, and efficient use of resources; hence the term *Intelligence Lead* policing. In today's Intelligence Lead police environment a police unit will move a few people into positions to deal with the criminals rather than just throwing officers at a problem. Often this is a proactive approach and it is increasingly in response to threats that have been identified through the intelligence process. The number of resources and scope of their duties would be chosen as a measured response to the criminal threat but it is no knee-jerk reaction. In general, this is how Criminal Intelligence is woven into modern police practice.

Nearly every police agency has a Criminal Intelligence analysis unit and at the Provincial, level they all work together pooling their information, which in turn is shared at the national level. Through analyzing specific information from the large pooling of police information, it is possible for criminal intelligence analysts to understand criminal behaviors and to predict future crime trends. Most serious police matters and organized crime targeting is highly dependent on Criminal Intelligence. In Alberta the provincial unit is called the Criminal Intelligence Service Alberta or C.I.S.A., for British Columbia its C.I.S.B.C. for Ontario its C.I.S.O. and so on. The Canadian body is called C.I.S.C., the Criminal Intelligence Service Canada.

For the most part, the general public is unaware of the idea of Criminal Intelligence concept and analysis. In its simplest form, Criminal Intelligence is amassing a great deal of information particular to one subject area or target, and determining what the information means, and working to suggest what may happen, or a course of action to be taken. It is really not that complex and ultimately this is something that people do all the time. Stockbrokers do a similar type of analysis to predict future trends in stock or commodities pricing. The meteorologist follows a similar process in predicting the weather.

What does this have to do with the criminal use of diamond? Statistics and history show that there is an increased criminal desire for jewellery and precious metals. By applying Criminal Intelligence analysis to this issue, the question 'why it is happening?' can be investigated. The cross-section of information that gives an insight into why there is a strong criminal desire for diamonds does not come from one single source or background. Several issues are interconnected. Factors that affect the appeal of diamonds to criminals include social issues, economic issues, national security, laws, and other national and international variables.

The following are factors affecting the criminal use of diamonds on a macro and micro scale. The macro factors deal with national or global issues that affect criminal use of diamond on a regional, national, or international scale. The macro factors may have a greater affect on the sophisticated or organized criminal. Micro factors involve issues that are more local in scope. These micro factors have a more direct impact on the low level street criminal, however, both micro and macro level factors are intimately linked as are the criminal actions of the organized criminal and the lowest street level criminal. **Macro Factors** will be discussed first.

Macro Factors

For organized criminals, criminals that are international in scope, or those people who are moving money in and out of countries, for example, to some off-shore account, cash can be a big problem in today's world. Nearly all the developed countries of the world have collectively agreed that organized crime and money laundering needs to be stopped. Rightfully so, depending on the source, it is estimated that international money laundering is valued at hundreds of billions of dollars per year. To deal with this, several countries have created financial security units that track large and/or suspicious movement of money within and between countries.

In Canada, the unit is called Fintrac (Financial Transaction Reports Analysis Centre), in the U.S. it's called Fincen (Financial Crimes Enforcement Centre), and in Australia it's called Austrac (Australian Transaction Reports Analysis Centre). These units are able to track the international movement of money. They are able to do this because many components of the financial services sector are required by law to report cash transactions in excess of $10,000. The units track these monetary transactions and they also have connections to the financial community's records.

To most people, $10, 000 seems like a lot of money. For drug dealers who are making three times this much money a week, it is not a lot of money. Rather, it is a problem. The question for them is how to hide the money from the authorities and other criminals. They can turn to money launderers, who for a substantial fee can perhaps help make these worries go away. Money launderers can take proceeds of crime and try to make it look like legitimate money for about 3%–25% of the costs. It's hard to keep the money in a bank or move it from account to account, or to transfer it out of the country without potential law enforcement intervention. It doesn't even have to be reported to be tracked, as the institutions keep all transaction records, whether it is $1 or $1,000,000 dollars. In fact, even countries like the Cayman Islands that use to be the safe havens for criminal money, are opening their financial sectors to the outside world[50]. Like the other countries, they also have dedicated units to track financial transactions. As a result, diamonds and gold can be a wonderful alternative to trusting a money launderer or the banks with criminal money. A single 1 carat diamond can easily be worth $5,000 U.S. at the wholesale level depending on quality. Keep in mind that for now I will talk wholesale prices for diamond because this is more relevant to criminal activities and this will be discussed in more detail in Chapter 8 "Diamonds as a Currency".

Unlike financial transactions, Fintrac does not track the buying and selling of diamonds, at least not yet in Canada, but this is coming. When they do start tracking these transactions, they will not be able to track the cash transactions that are so prevalent in the industry. In recent years, post 9/11, international efforts to uncover terrorist accounts and financial sources have further corralled criminals into these other non-cash commodities. Because of this, some governments have recognized the criminal use of diamonds and have legislated that transactions of diamond purchases over $10,000 will also be reported. Thus far, this legislation is found in very few countries around the world. However, regardless of where this legislation exists or where it may be forthcoming, the very nature of diamond, diamond valuations, and the diamond business makes non-compliance for criminals very easy. Many have written about the traditions within the diamond business, and the cash and handshake tradition of the diamond industry[51]. This tradition is alive and well and certainly would make transaction tracking difficult. The subjectivity of diamond valuations is another major impediment to effective legislation respecting diamond transactions.

Diamonds are not captured on financial reports like cash that is easily tracked by computer, however they may, at times, appear on inventories. Yet, using a transaction reporting thresholds of $10,000, who would know if diamonds worth $30,000 that are sold to a criminal or terrorist are falsely valued at $9,000? The cash transaction is less that $10,000 and not reportable unless the vendor has a suspicion that the transaction is liked to money laundering. There is ambiguity in defining what is suspicious and a criminal could sell diamonds to other criminals in this manner as a money laundering service. Once the stone is sold and is out the door, the authorities cannot go back to check on this. It is likely the authorities would not even know about any stone that goes out the door under $10,000 dollars. No report, no buyers name, nothing.

To further legitimize this, Criminals can pad their inventories by purchasing stolen diamonds from other criminals off the street or through acquisitions from pawn shops and auctions. Diamonds that are not on inventory can be sold by the criminal at any price entirely under the table and off the radar of the authorities. Diamonds present one of the greatest opportunities for manipulation of inventories, and more importantly capital and values.[52] Then again, even if there are inventory records, it would be difficult to perform an audit on all ledgers of a business while cross-referencing the book values to the actual values of the inventory and off-inventory stones. A complete and thorough audit/reconciliation process requires a multidisciplinary approach of investigators, accountants, and gemologists. Depending on the investigation it may be difficult to determine if there even is a potential infraction. Of course no system is foolproof but as discussed, the very nature of diamond valuation is quite exploitable by criminals. The bottom line is that legislation focused on diamond sales transactions, whether at the retail jewellery level or higher, is difficult to draft without loopholes, and probably more difficult to enforce.

The Grading and Valuation of Diamonds

Valuations are discussed in Chapter 3, but to be more direct, the subjectivity of diamond grading and the compact value of diamonds are exploitable points for criminal use of diamond. In general, each of the colour, clarity, and cut characteristics of a diamond are no more absolute than saying the sky is blue. In fact, the sky is the colour blue, but if asked to describe the colour without saying "It is sky blue", it would be hard to answer. One can say powder blue or milky blue, baby blue, or sea blue. In the end, each individual will see it differently. Grading diamonds is the same. For all those but the most experienced diamond graders, the subjectivity allows for grades to swing wildly from grader to grader on the same stone. This is the foundation for criminal deception and exploitation because the grading variation can be someone's honest belief of a diamond's grade or the grade assigned to the diamond can be a deliberate deception. In a recent study, 13 diamonds were taken from one source and these diamonds were sent to three of the top labs in North America, American Gemological Society (A.G.S.), Gemological Institute of America (G.I.A.), and European Gemological Laboratory (E.G.L.). These are the labs with the experts, with the diamond graders who are considered by many as the best of the best, and who adhere to the most rigid of standards. Not one diamond got the same grade from all three labs and several of them received different grades from each laboratory[53]. Despite the different grades received for each diamond, it was surprising to see just how close the valuations were. When I read the study I expected the valuations to vary more greatly than they did, so kudos to A.G.S., G.I.A., and E.G.L. as well as Pricescope for conducting and publishing the study.

The reality is that even at this top level, the valuations were different, and justifiably so. Because each of the clarity, colour, and cut of a diamond are grading continuums with no precise boundaries. Depending on the criminal activity, it is easy to over-grade or under-grade a diamond in any one of these valuation categories to suit a purpose. In its simplest form, over-graded diamonds can be used to facilitate money laundering and frauds and under-graded diamonds can be used for tax evasion purposes.

Universality of Diamond

In addition to characteristics that are easily exploited by criminals, diamonds are universally sought after by criminals on all continents. As such the ability for someone to buy, sell or trade diamonds throughout the world makes this item very appealing to criminals—especially organized international criminals. This probably isn't so for absolutely every country as no doubt there are underdeveloped countries of the world that really don't care about diamonds. Tibet does not have the diamond buying and selling markets as in North America or in Europe and diamonds may not have the same allure there as they do here.

The ability and speed with which one can turn a non-cash asset into cash is what bankers refer to as the liquidity of an asset. A house, for instance, is not very liquid as it requires a great deal of paperwork and movement of money through lawyers over what is usually several days or weeks to complete a transaction. This is assuming a buyer is found for the house right away. On the other hand, diamonds can be sold in a major metropolitan center within 24 hours[54]. As a result, its rating as a liquid asset is high. This is another great quality that diamond has. A criminal that transports diamonds from Australia can easily sell them in Canada, the United States, Great Britain, South Africa, or Europe. In addition, when dealing with criminal entities in other countries, the universality of diamond makes it an excellent choice for paying for drugs, weapons, or other contraband. Equally important is an international pricing scheme, which makes holding diamonds in a country with an unstable currency better than holding cash. Saddam Hussein was reported to have purchased large quantities of diamond jewellery just prior to the 2003 coalition invasion of Iraq as a means of hiding money and stabilizing his squandered wealth[55]. In fact, the paper money that was used while he was in power has now been replaced with a totally new currency while diamonds purchased remain a valuable cash-like instrument and stable in an unstable environment; perfect for any fleeing dictator.

Diamonds have proven to be stable in times of economic turmoil, and have increased in value over time and at an equal or greater rate than inflation[56]. The other great quality of diamond as an asset for criminals is that it is virtually untraceable. Unlike property, vehicles, cash, and nearly anything else of value, there is no registration, no title of ownership, and with few exceptions, no serial numbers, although some companies are now laser marking their diamonds with serial numbers. Even exceptional diamonds that are undoubtedly identifiable, like the French Blue, can be recut or polished to remove or change any identifying features.

The addition of a serial number to a diamond for loss prevention or security measures doesn't seem to help significantly for recovering stolen diamonds nor act as a deterrent for theft. Use of serial numbers on diamonds is very rare in the grand scheme of things, but on Canadian certified mined diamonds, it is a requirement. The serial number is laser etched onto the girdle of the diamond so as to not detract from the brilliance or clarity of the stone. The laser serial number is so small that usually it can only be seen on the girdle of the diamond through 10x magnification. Also placing the serial number on the girdle makes for a universally convenient place to look for the marking. In terms of theft, the serial number is only valuable if it is recorded or the certificate with the number on it can be located. Then if the diamond is stolen, the serial number may become valuable as an identification tool. A jeweller who comes across a diamond with a serial number and no certificate may be suspicious and decide not to buy it or deal with it at all. This was the case when a shipment of diamonds transported by airplane from Edmonton to Vancouver went missing enroute. Several of the stones turned up later when the criminals tried selling them to a Vancouver jeweller who noticed the serial numbers, yet no certificates accompanied the diamonds. The jeweller tipped off the police and the criminal was arrested. On the other hand if the diamond is offered up at an incredible price a criminal operating as a jeweller may decide to buy it and have the diamond mounted in a piece of jewellery that conceals or hides the girdle and serial number. If the diamonds are large stones, they could be spot polished or recut to remove the serial number. Considering the number of Canadian diamonds stolen with serial numbers, there are no statistics that show the recovery rates of these stones are better than diamonds without serial numbers.

Globalization of Business and Criminals

In the past 20 years, and even more so with the evolution of the Internet and travel, the old cliché that the world has gotten smaller and the business world has gotten bigger, could not be more true. Companies that once were regional players or even national entities have become international in scope by reaching out to new business opportunities. Companies that have not kept up are now just getting by, may have a unique market, may have been swallowed up by the market expansion, or have formed strategic alliances with other companies. Criminal operations are very much like those of the business world in that they expand and evolve to take advantage of new market opportunities. The Hells Angels, once an organization that existed only in the United States, have now expanded to nearly all continents—an export that no country wanted. Likewise, Italian Organized Crime has expanded from Europe into all regions of the world as have Asian organized criminals. The most recent expansion has been of the East European Organized Criminals who have exploded through all continents since the fall of communism in Eastern Europe.

The criminals are much more effective at moving into an area than companies in the business world for many reasons. First, they often don't have any competition and if they do, like the business world, one company is so much bigger than the other and simply swallows the other up. In the criminal world of Outlaw Motorcycle Gangs (OMG's), this is what is known as a "patch over". A "patch over" happens when one motorcycle group decides to amalgamate with another. It is always the larger more powerful OMG that wins out. OMG's are very territorial and often a "patch over" doesn't come easy or at all, and this results in what could be described tongue-in-cheek as a 'hostile takeover'. In Canada, some of the most violent actions have occurred in Montreal during the war between the Hells Angels and the Rock Machine from Quebec. Other OMG's like the Banditos of Edmonton simply rolled over to the Hells Angels. The "patch over" of the Edmonton Banditos to the Hells Angels came quickly after one of the Banditos was shot to death in a yet unsolved homicide.

Typically each organized group concentrates on one or two lead brands of criminal activity in which they excel. They may dabble in some other activity but it is not the focus and for certain, some other organized criminal group somewhere else is doing it bigger and better. Also, there are efficiencies that come with making decisions in a criminal organization. There is no concern with the environment, the public, health and safety, or the law. There are no board of directors, no shareholders to answer to, the decisions are made quickly, decisively and efficiently. When these issues don't have to be taken into consideration, it is very easy to find a quick route to profits. A similarity the criminal world has with the business world is working alliances. This has not always been the case but as organized crime becomes more sophisticated and expands its borders it begins to form business like arrangements with other O.C. groups[57]. The business arrangements can include supplying a particular type of drug, firearms, or money laundering services.

It is expected that in the world of diamonds, some groups are far better at using them than other groups. A criminal group that operates extensively using diamonds can demand diamonds as alternate forms of payment for contraband that has been supplied. In other areas of the world, East European Organized Criminals have operated for decades in West Africa, South America, and the former Soviet Union. These criminals have incorporated diamonds into their criminal activities in Canada[58]. Diamonds are a commodity that can be bought or sold throughout the world. This means that diamonds are an alternative currency that criminals can utilize for conducting business activities while avoiding the probing government agencies that are watching the international movements of money. Increasing the scope and reach of the business activities of organized crime requires new ways to do business. They have a need to develop new and innovative ways to move contraband and profits country to country undetected by authorities. In this respect diamonds have been used historically, however, as international security measures and money tracking initiatives expand globally, it is expected that the criminal use of diamonds will also expand and at an ever-increasing pace.

Proceeds of Crime and Money Laundering

In the early 1900's, Arnold Rothstein, a grand daddy of Organized Crime in North America, was already using stolen diamonds within his criminal organization[59]. This being the case it would seem likely that DALC could be incorporated into a criminal organization for the purpose of money laundering. Yet globally these commodities were not seen as a threat for money laundering until 1998, when they first appear as a mention on the FATF annual reports[60]. However, even in the big picture there is little information about diamonds being used widely among criminal organizations until the past two decades but even that data is limited. One analysis conducted suggests that of the proceeds of crime cases investigated by the R.C.M.P. in the early 1990's, 10 percent of the cases involved police seizures of diamonds and like commodities[61]. Yet, more recent data suggest that the criminal use of these commodities is accelerating quickly. Until recently if criminals were making large amounts of money on illegal liquor, drugs or other illicit endeavors, they could utilize the services of off-shore banking to hide their proceeds of crime and launder money. They would not only be able to keep it safe, but also be able to use the money at some time in the future without drawing the attention of the government, law enforcement, and tax departments. Cuba, Switzerland, the Bahamas and Cayman Islands, among several other countries, were all at one time havens for criminals to both hide money and or launder it, often through casinos[62]. Many of these countries allowed for the creation of privately owned banks and rules that not only promoted but ensured that individuals could secretly hold a bank account by number only without anyone knowing his name, hence the name—numbered bank account.

The practice of anonymity through numbered bank accounts is beginning to go by the wayside due to the international pressures exerted by powerful countries to try to stop the international laundering of money. Of course, the only reason a criminal needs to launder money is if the money is from proceeds of crime. That being the case, the United Nations estimates that international money laundering figures to be between 500 billion and 1 trillion dollars[63]. This represents 'dirty money' of a purely criminal origin. The costs go beyond the figures presented, as this is simply the profit generated from crime. Associated to these figures are the millions of lives ruined and persons victimized through the criminal activity that generated these monies. This is not the only cost, it is estimated that the United States alone spends over $600 million annually simply fighting money laundering, and those costs are rising[64].

What is a criminal to do if he can't hide the money in a bank? He has millions of dollars but can't show that he has worked an honest day in his life. He could put it under his mattress, but as soon as some other criminal finds out he is doing that, he will likely get bumped off for the money. The other problem is a million dollars worth of U.S. $1 bills, if nicely packed, would still take up a lot of room, approximately 42 cubic feet, and would weight 1 ton.[65] The criminal could buy a lot of nice property, houses, and fancy cars, but these items are easy to track by the authorities. He could buy a business and try to use it as a facility to launder the rest of his money. This is not a bad idea but again the Proceeds of Crime Units and forensic accountants are brilliant at uncovering the criminal activity in these actions. He could try to hide it in a privately held foreign bank beyond the reach of authorities in North America, but recovering the money is not fool proof. Hiding money is not as easy as it used to be and criminals are turning to diamonds and like commodities to hide and convert money. These like commodities are precious gemstone and precious metals. In particular the diamond trade has been singled out as an industry that is vulnerable to money laundering as are jurisdictions that have significant trade in or export of gold, diamonds and other gems.[66]

It is not only criminals, but it is believed terrorists would also use diamonds and like commodities as a means to finance their operations[67]. As for diamonds, the criminal purchase or acquisition of diamonds allows them to dump a huge amount of money into a product that in short order is easily converted back into cash. Approximately eight, 1.00 carat high quality diamonds are worth approximately $100,000 at the wholesale level. These stones would fill about half a thimble and the best part for the criminal is that the authorities don't know who has them. There is no way for law enforcement to track the diamonds or to know that the criminal has them unless by chance they stumble across them. The police might confiscate the Porsche as proceeds of crime, but the diamonds are not exactly the kind of thing they'd find records of or easily trip over during a raid. Selling the diamonds is easy in North American metropolitan areas and like the purchase of the diamonds, there is no tracking of who sold the diamond. The criminal simply has to sell five good stones and he's driving a great car again.

The ability to buy and sell diamonds without the authorities knowing is very similar to what has been done in the past with cash deposits and withdrawals to private banks. Diamonds can be the vehicle for acquisition of wealth and for the storage and movement of money. In addition, diamonds have historically gained in value over the years, are seen by some as an investment, and as a hedge against inflation.[68] As a cash-like instrument, the same eight high quality diamonds sewn into the cuff of a pair of pants are easily taken anywhere in the world. The diamonds can also be placed into the cavity of a personal computer or electronics device as they will not short circuit the device nor show up on x-rays. There are so many ways to move diamonds surreptitiously, and there are even books written about this[69]. Diamonds are not detected by most x-ray machines, nor are they detected by metal detectors or police dogs. Diamonds increase in value over time and in the hands of a criminal, in essence he becomes his own private banker. Purchasing or otherwise acquiring diamonds becomes a deposit, while selling diamonds or trading them is like a withdrawal. The hiding of wealth through the purchase of diamonds doesn't replace the criminals' need to show the legitimacy of the wealth. However, the diamonds themselves, unless they are displayed, would not draw any attention to a criminal.

Hypothetically, the brazen criminal could actually take a parcel of diamonds to the police and say he found them. The police would have to report to the public that a parcel of diamonds was found in order to try to locate the owner. The police could say there were several stones and the parcel is quite valuable but no specific details of the parcel could be released. This would ensure to the police that whoever the supposed real owner of the diamonds was would have to produce the complete details on the parcel in order to positively identify it. Of course, no one would be able to step up and positively identify the parcel because they were the criminal's diamonds in the first place. After a short period of trying to find the owner with no success, the police would be obliged to give the diamonds back to the criminal—finders keepers. In that process the diamonds would have been legitimized and laundered.

In real life, other schemes that do not involve or draw the attention of the police are employed and the subjective value of diamond can be manipulated for money laundering purposes. As such diamonds can be incorporated into any number of businesses within the jewellery industry from manufacturer to retailer, and through over-valuation the diamonds become a medium for money laundering. This will be discussed in detail in the next chapter—Criminal Use of Diamond. Suffice to say that diamonds are very appealing to those criminals with the 'problem' of concealing their criminal profits from authorities.

Diamond Specific Laws

In Chapter 4, diamond laws in Canada were discussed, and relatively speaking it is a short chapter. The bottom line is there are really no laws that are specific to finished diamonds in Canada or for that matter in the United States aside from the Export and Import of Rough Diamonds Act (Canada) and The Clean Diamond Act (U.S.). This affords criminals an excellent opportunity to use a commodity that has little or no controls on it. In virtually every enterprise that criminals are involved in, there are controls or are laws prohibiting the activity. Whether it is prostitution, drugs, cigarettes, counterfeit goods, or firearms, law enforcement has some means of dismantling the criminal network when it has been uncovered. Even in secondary activities dealing with the proceeds of crime and money laundering, police have ways to charge and seize assets through specific legislation and units designed to detect these activities. Criminals can operate with diamonds as a means to criminal financing, for money laundering and for hiding and moving money. Although, Proceeds of Crime (P.O.C) and money laundering legislation can cover diamonds and like commodities, it is not always easy to tie the diamonds into the criminal organization or the persons involved. Ownership and acquisition of the diamonds is an issue because there are no title documents, transaction details are difficult to come by, and stolen or illicit diamonds are very difficult to trace or track.

The jewellery industry itself is particularly unregulated and although I'm not in the habit of advocating new laws or regulations, the lack thereof is an open door for criminal activity to move into and participate in the industry. Countries like South Africa and Belgium have laws requiring the registration and licensing of diamond dealers and cutters. Having said that, even the laws, regulations and over 400 police officers dedicated to diamond and precious metals crimes doesn't extinguish the criminal activity in South Africa. The creation of diamond and/or jewellery specific laws may not stop the criminal activity, but it does give law enforcement some tools to address the problems. The lack of laws or regulatory controls in countries with a strong and vibrant diamond industry certainly increases the potential for criminal exploitation of diamonds.

Unknown and Unchecked by Law Enforcement

Douglas Farah, author and investigative reporter for the Washington Post, is quoted saying that there has been a very limited interest by intelligence and law enforcement in terrorist financing through diamonds[70]. I agree with Farah, and submit that the assertion could be extended to also include the broader scope of criminal use of diamonds and like commodities in general. Currently, in Canada, diamonds and jewellery fall under the domain of many law enforcement agencies including the Royal Canadian Mounted Police (R.C.M.P.), Municipal Police Agencies, Natural Resources Canada, Industry Canada, Heritage Canada, and Canadian Border Services Agency among others. That being the case it is expected that within these agencies there should be a compendium of knowledge and experience specific to diamonds and jewellery crime. This is the case, except that the totality of knowledge and experience is stretched over so many officers throughout so many units, it is too watered down in any one unit to be effective. The exceptions are those units that are dedicated to diamond and jewellery crimes.

Within the R.C.M.P, several units have the potential to deal regularly with diamond and jewellery related criminal activity. Those units are the Proceeds of Crime Units (P.O.C.), Commercial Crime (C.C.), the Integrated Market Enforcement Team (I.M.E.T.), the Customs and Excise Units (C&E), Drug Units, General Investigation Sections (G.I.S.), and General Duty Units (G.D.). Proceeds of Crime seize diamonds and jewellery as a regular function of their duties, the Commercial Crime Unit investigate businesses that are utilizing diamond as a part of their criminal activities, Customs and Excise Units investigate the smuggling of diamonds, G.I.S. investigates jewellery store robberies, and G.D investigates business and residential break and enters where jewellery is the target of criminal activity. Despite the fact that so many of these officers deal with events involving diamonds and diamond jewellery, most officers have very little knowledge of this subject. 'Diamonds 101' is not a training module at the RCMP training academy or any other police academy. With criminal use of diamonds on the rise, the prospect of terrorist use of diamonds, and Canada currently the third largest producer of diamonds, Canadian police colleges may soon have diamonds or jewellery industry crime on the curriculum.

The average officer that comes out of a training academy knows the laws with respect to diamonds. Then again, that is because there really aren't any. With drugs there are some specific laws and these are required materials which the officers generally have a good handle on when they leave the academy. What most officers do not have is the experience to spot the drug users and what to look for as indicators. Unless the officer experienced more than academics while attending university or college, this knowledge is something that is typically passed on from senior investigators down to the junior ones and is a function of on the job training.

Knowing how to deal with diamonds and diamond jewellery is a critical first step in any such investigation. When a break and enter occurs, the victim is asked to supply a list of what was stolen and the police agency enters on its log of stolen property. Occasionally the victim requires assistance completing this and the officer does what he can to help compile the list. In any event, when it comes to diamonds and like commodities, what usually results is a list something like this;

Stolen Item #1) gold chain with heart shaped diamond pendant

Stolen Item #2) mens gold wedding band with a diamond

Stolen Item #3) ladies gold hoop earrings with diamonds

The description of these items could account for dozens of completely different items within a jewellery store. Even if an officer came across these items through the arrest of a criminal, what exactly could be turned back over to the victim? Is there anything that can positively identify the items as belonging to the victim? For instance how long is the chain and what kind of linkage assembly is it, what exactly does the heart shaped pendant look like? Does the chain feed through an opening in the heart or is there a bail that suspends the heart pendant from the chain? What karat is the gold? What size is the men's wedding band? How wide is it? Is it a regular band or comfort fit? Are there any particular markings or fine gold work? A complete description of these items can be nearly irrefutable and allow for easy identification especially through other police records or pawnshop records.

The majority of this information is the sole responsibility of the person who owns the jewellery. The description supplied is as good as the victim's recollection and also as good as the officer's knowledge of jewellery and ability then to ask questions to elicit the most comprehensive description. In this respect, simply asking the right questions to obtain a complete description may lead to a higher chance of recovery. What the officers don't have though, at the street level or throughout most of the units, is a basic knowledge of jewellery beyond that of the average person. Perhaps they bought a large diamond once, but I would suggest that 4 out of 5 people that purchased a large diamond at some time in their life could not with certainty state the four C's of that stone. For anecdotal purposes, I have asked this question at workshops and seminars and the results are nearly always the same. Only once have I had someone that could fully describe the characteristics (Four C's) of the diamond they had purchased.

The problem continues when property is recovered or property is turned into the police agencies' lost and found department. The jewellery inevitably will be described in the most innocuous terms so as not to put an undue value on the item. For example, an officer recovers a small stash of jewellery from a criminal that the officer believes is stolen. There are three pieces of jewellery, each with a gemstone. The descriptions could be as follows:

Recovered Item #1) 16" yellow chain with pendant and large white stone

Recovered Item #2) ring with large clear white stone

Recovered Item #3) ring with large oval white stone

The officer can not describe the jewellery in any more detail than this because of future implications. For instance if he describes Item#1 as having a diamond in it and for whatever reason the criminal gets out of his charges and wants the property back, he is going to be expecting—a 16" yellow chain with a pendant and large diamond. If the stone was really a cubic zirconium, but the officer said it was a diamond, the criminal can make a claim against the police to get what the paperwork says it is. What this means is that the jewellery recovered is often so poorly described that it cannot be married up with what may show up as being stolen. The limited description is to protect the organization taking the items in. Item#1 may be well described in a loss report as a cubic zirconium but this won't necessarily cross reference as a white stone, nor will it query as such through police records or other data banks. In fact, querying gemstones was particularly difficult until recently, as there was no way to enter a gemstone into electronic records by its carat weight. Until recently, the stone's carat weight had to be described in a narrative that could not be queried through any search function. The Jewellers Vigilance of Canada assisted the RCMP in rectifying this problem for data entry of carat weight and now a stone's carat weight can be added to a description field.

What this all means is that beyond the inherent properties of diamonds and jewellery that make them difficult to trace and track, the limited police and public ability to describe and record stolen diamond and jewellery items exacerbates the issue and no doubt contributes to the low police recovery rates of jewellery and precious metals[71]. It's important to note that the lack of knowledge of diamond and diamond jewellery by the public is an important factor in the limitations that police have in tracing stolen diamonds and jewellery. In fact, historically the public has always known very little about all aspects of diamond, even among those who mined diamonds in the late 1800's. One of the tests that early miners use to employ to determine if the stone they had found was diamond or not was to subject it to the blow of a sledgehammer[72]. The idea behind this test was that if diamonds are supposed to be the hardest known substance then it can withstand a blow from a sledgehammer. Diamonds are hard, but not necessarily that tough and one has to wonder how many diamonds were reduced to dust at the hands of an unknowing diamond miner.

As organizations, law enforcement is not well versed in the criminal use of diamonds or of jewellery industry knowledge[73]. Few statistics are kept or compiled on these matters and much of the information obtained comes from open sources. Open sources, are sources of information beyond the classified police data and includes all publicly accessible data sources. This is a great method of obtaining information but as any intelligence officer knows it can be difficult to verify and qualify the information. In addition to knowledge gaps respecting the amount of criminal activity related to diamonds and diamond jewellery, the police have no specific laws aimed at tackling the growing criminal use of diamond.

Police often measure their effectiveness by the number of offences solved and will dedicate resources in order to stymie crime where it appears. Police policy and strategy is largely based on enforcing laws drafted by government. As it pertains to diamonds, there are virtually no laws and no specific criminal action that can be targeted by police. As such, most law enforcement agencies have little policy for diamond specific investigations. This further translates into a limitation on resources dedicated to diamond specific crime. In so far as criminals are concerned, this is their oasis in a law enforcement desert. There are no diamond specific laws, and law enforcement agencies rarely even look at these crimes specifically, but instead as incidental to other criminal actions. For every other widespread and serious criminal activity, there are officers and units ready to take down the criminals except in relation to diamonds specific criminal activity.

Recently law enforcement has begun to recognize the criminal threat and law enforcement vacuum that surrounds diamond crime. In August 2004, the Canadian Associations of Chiefs of Police through the release of the Criminal Intelligence Service Canada report on Organized Crime identified Organized Crime as a threat to the diamond industry[74]. This is a step in the right direction. The R.C.M.P. recognized the potential threat to the emerging diamond production industry long ago. In 1995, three years before the first diamond produced in Canada, the R.C.M.P. Diamond Protection Service was established in Yellowknife, Northwest Territories. There are other units and programs emerging; in Ontario, the RCMP has established a jewellery unit that investigates criminal activities related to jewellery. In the United States, the F.B.I. has set up a Jewelry and Gemstone Unit (J.A.G.) that focuses on jewellery industry crimes and they have partnered up with the jewellery industry to stop organized criminal attacks on jewellery stores and traveling jewellers that has plagued much of the United States in recent years. These are all steps in the right direction to address the present criminal activities and emerging threats.

North American Diamond Culture

For centuries, the people of India cherished gold as a method of storing wealth. It was a convenient and smart way of keeping one's wealth close at hand in the event of having to flee due to marauding invaders or a natural disaster. But to Indians, gold as a symbol of wealth and status is interwoven into their society such that it could be said they have a *gold culture*. In the same respect, a similar label could be applied to North Americans in their desire for diamonds and be called a *diamond culture*. Diamonds are produced and finished the world over and in spite of the cutting business in the United States and Canada, it really is just a small part of the overall world diamond industry. On the other hand, countries like Belgium, Israel, and India cut far more diamonds per capita and have a significantly greater flow of the stones through their borders than in North America, yet when it comes to a desire for diamonds, no one tops North America. To be clear, it the U.S. market that must be discussed. The best estimates peg the annual world diamond jewellery market at approximately $60 billion U.S. and of that, the United States consumes nearly half of that production[75]. That is about 30 billion dollars worth of diamond jewellery consumed by approximately 6% of the world's population.

North Americans are extremely wealthy compared to many other countries of the world and that helps fuel jewellery and diamond sales but even those on the bottom rung of the economic ladder have diamond jewellery. North American pop culture is riddled with more examples of their love for diamonds and jewellery. The Academy Awards inevitably host some starlet who is wearing something borrowed from Harry Winston, 'The Jeweller to the Stars'. Sport stars are adorned with magnificent rings upon winning a championship title. Music icons, with a nod to the rap artist, are laden with pounds of gold chain. Diamonds and jewellery are woven intimately into Western culture and similar to India in that respect, possession of these objects represents wealth and status. In order to quench the North American thirst for jewellery the jewellery market has evolved to accommodate this. Internet jewellery sales, television shopping, and a jewellery store on every street are all fully supported by the market forces, and demand is growing. Again, further up the mine to market chain, there is that many more diamond and jewellery importers, manufacturers, dealers, and wholesalers than one would find anywhere else. What this means is that in terms of criminal acquisition of diamond and jewellery, there is no better place in the world to get hold of diamonds and jewellery. Whether through a break and enter, home invasion, jewellery store theft or robbery, a criminal has easy access to diamond jewellery virtually anywhere in North America.

Obtaining diamond jewellery is one thing, but once criminals have these items they have to dispose of them in order to reap the benefits of their labours. In this respect, the same market conditions that allow criminals to easily obtain diamond jewellery also allows them to easily dispose of the jewellery. Pawn shops, jewellery stores, diamond dealers, auctions, and the internet provide plenty of opportunity for criminals to sell the diamond jewellery by inserting the stolen jewellery back into the jewellery market.

Diamonds are Good Business

There is good money to be made on diamonds, especially at the retail level, but that's not to say this business is profitable for everyone. The positive outlook for the diamond industry could be expected to continue for some time into the future. Currently, the world demand for diamond is outpacing supply. Natural diamonds are a finite commodity, and as such, with the foregoing supply/demand trend, the value of diamonds and expected profit margins can be expected to increase. The margins are significantly lower at the diamond wholesalers or dealer level, but this is offset by the number of retail outlets to which they are selling. The profits are made in bulk sales rather than individual stone sales. If one jewellery store isn't selling much, there are usually others that are selling plenty of diamonds. As long as diamonds are supplied to the busy stores, profits will be good. That is a great place to do business and this end of the industry is a good place for criminals to move in to the industry.

While diamonds are good business, when used by criminals, they present a good front for illegal activities and an opportunity to facilitate criminal activity directly through the use of diamonds. In South Africa, where diamond dealing is controlled by law, police have raided Organized Crime rings that deal both in drugs and illicit diamonds.[76] A criminal investment of money into a diamond dealing operation or the likes somewhere at the intermediate level of the mine to market chain allows them to open a business from which to generate legal money but also affords them the opportunity to engage in money laundering activities. With the international demand for diamond increasing, the appeal of this industry to criminals is even greater. The fact that there are few industry regulations or controls across North America sweetens the pot.

This doesn't mean that there are numerous criminals operating at this level of the industry. On the contrary, the people in the industry understand the criminal potential and through the Jewellers Vigilance Canada and the Canadian Jewellers Association and other industry advocates, steps are being taken to thwart criminal interventions. However, this is an industry that is profitable and all indications are that it will continue to be profitable and even more so into the future. Criminals operating at this level recognize this and for these and other reasons, are eager to get into the industry for the profit and criminal potential, while utilizing the industry as cover for their illegal operations.

Micro Factors

Coincidental to the diamond specific criminal activity that is occurring at the national and global levels is the increasing diamond specific criminal activity that is occurring at local and regional levels. This is evident in the statistics that have emerged from the United States clearly showing that a criminal desire for these products exists. It is apparent that this street level appeal of diamonds to criminals at the 'micro' level is closely linked with appeal to criminals at the 'macro' level. To understand why criminals are so interested in diamonds and jewellery, it is necessary to examine some of the issues that affect the criminal world as well as society in general. Technology, economics, and social change play a role in how a criminal can operate and although no direct data or studies exist that examine the totality of this issue, there are several changes that have emerged in the past 10–15 years that make diamond and jewellery more desirable to criminals.

Increasing Values of Diamond and Jewellery

My mother has a beautiful diamond engagement ring that was given to her by my father over 40 years ago. It has a center diamond of about 0.75 points and is set in 14 karat yellow gold. That ring was purchased for about $300 at that time. Today the retail value of that ring is approximately $5500. The value has increased nearly 2000% in that time, which also amounts to about a 7% per year increase in value. Imagine getting 7% interest, year after year for money in the bank. This is one of the greatest characteristics of diamonds and precious metals in that it typically increases with inflation. So when a diamond or piece of jewellery is stolen, its value is considered as new, as if it were stolen directly from a jewellery store that day. This holds true for diamonds and jewellery regardless if it is stolen from a residence in a break and enter or through a personal robbery. Aside from a monetary instrument, there are few other items that carry these increasing values through time. Even monetary instruments carry risk of devaluation or becoming valueless. This was seen with the pre-World War II German Mark that was replaced after the war and the Iraqi dollar under the Hussein regime that has been replaced with a new Iraqi dollar during the U.S. led war. In addition, what a dollar could buy 40 years ago is only a fraction of what it can buy today due to inflation. Holding diamonds for wealth may be better than having money in the bank and this is why. If $1 is put in the bank at today's daily savings account interest rate of 1–2%, and given the annual inflation rate at over 2%, in one (1) year from now that same dollar has less buying power than it does today. On the other hand, historically diamonds have increased in pace with or at a greater rate than the annual inflation rate[77]. This means that regardless of what diamonds or diamond jewellery a criminal acquires, these items will maintain the integrity of their value more so than money in the bank. For criminals that don't need to convert the stolen diamonds or jewellery to cash immediately, this means they can hold on to these items without the risk of them losing value over time. In fact, as we know, they will increase in value over the long term. For criminals that convert the acquired diamonds and jewellery to cash, they realize that can obtain top dollar for these items regardless of where they were obtained and how old they are.

Easily Converted to Cash

I was in a large pawn shop in the late fall of 2003 checking out the unique jewellery pieces they had for sale. While at the pawn shop the clerk was negotiating with a customer on a diamond ring that the customer wanted to pawn. The man asked how much he could get for the ring and the clerk was checking the size of the stone to estimate its weight. While this transaction was going on another man came in off the street pushing this bicycle. This bike looked great, the latest in road equipment and was probably worth $5000 or more. My guess was that this fellow might get $200 for the bike, less than 5% of its value. To my surprise, the clerk saw the man coming in with the bike and before the man could even say anything, the clerk said "We're not taking any bikes right now. Come back in the spring". This turned out to be an opportune moment to see how easy it was to convert diamonds to cash versus other commodities. Here was the first man with a diamond ring that the pawn shop had no problem paying cash for, yet the second man with a fantastic bike couldn't even get an offer. To finish the story, the clerk offered the man with the diamond about $185 for the ring. He turned it down, saying that the diamond weighed more than the weight the clerk had established for the diamond. The man with the diamond ring left the pawn shop and went into the other pawn shop across the street. No doubt he will get a better price for the ring if he checks out a few other pawn shops or jewellery stores.

Pawn shops are more than happy to take the rings off peoples' hands—no pun intended. There is probably no better place than Reno, Nevada to see just how convertible diamonds and jewellery are through pawn shops. In Reno's city center there are a couple of very large casinos that cover the entirety of approximately two city blocks. On the other side of the street are two blocks of stores, nearly all of them pawn shops. These pawn shops are conveniently located so that when a gambler runs out of money, he can simply walk across the street, pawn his diamond ring or any other jewellery he has and head back to the casino with the money on loan. In fact, some of the pawn shops are exclusive to jewellery, and will not purchase anything else. There is nothing wrong with this and it is common all over North America. Simply look at the yellow pages under jewellers or pawn shops for the Toronto market and you'll find a list of pawn shops or jewellers advertising to buy jewellery off the street.

Pawn shops are not the only places that diamonds and jewellery are sold and sold easily. Bill Mason writes in his autobiography "Confessions of a Master Jewel Thief" about how jewellers and auction houses, even world famous auction houses, were more than happy to purchase the jewellery from him or in the case of auctions, put it up for auction on his behalf so they could collect the commissions[78]. According to an article in Jewelers Circular Keystone Magazine, jewellers suggest that auction houses overlook forgeries and forget jewellery disclosure rules[79]. In fact, the same article cited statistics showing a dramatic four fold increase in the value of jewellery being sold through high-end auction houses between 1987 and 1997. The moral of the story is that diamonds and diamond jewellery are easily turned into cash. For a criminal looking to turn a quick buck on some product he can steal, given the choice between some jewellery and a bike, he is going for the jewellery. This leads into the next point.

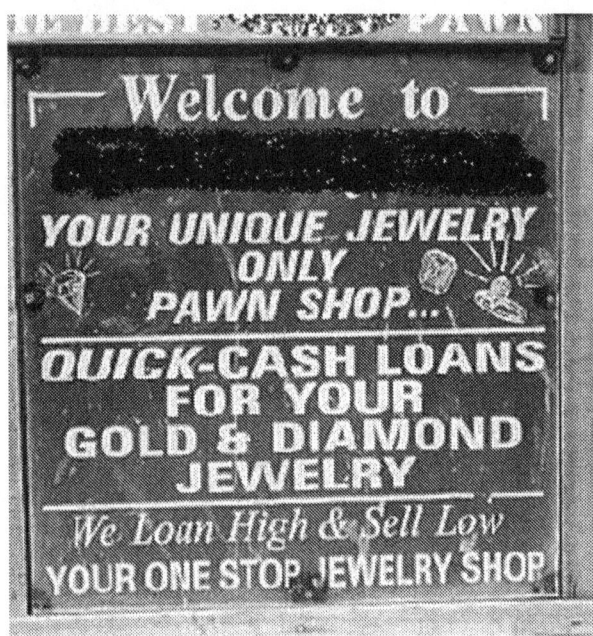

Easy to Conceal, Carry, and Store

The streetwise criminal that does a break and enter into a residence is well aware of the characteristics of property that make them easy to detect by the police. One is the size if the property. If the items are not easily hidden, that can present a problem for the thief. A bike can present a problem to a criminal who commits a break and enter and then tries to leave the premises with the bike. First if the criminals are driving a car, there is a good chance they would not even be able to get the bike into the car. If a criminal is on foot, then the bike can make a good get-away vehicle, but anyone who sees the criminal with the bike is now a potential witness for the police. Unlike jewellery, a bike can't be slipped into a pocket or carried along without someone seeing it. On the other hand, as one very prominent person discovered, a ring can be slipped into a pocket, but this act may be caught on camera[80].

There is also the problem of storage. A career criminal who steals on a regular basis may accumulate a large quantity of property over a relatively short period of time. If the criminal waits for the investigational fervor to wear off before turning over the property for cash or drugs, he will need to store it for some time until he does so. Large property items can cause storage space problems. This is especially so for travelling criminals and criminal groups. If the criminal is a fence, someone who acts as a middle-man between the thief and the pawn shop or other legitimate sales outlet, he may accumulate an even greater amount of property, which again can cause a storage space problem. The storage problems arise not only in terms of space but also in being able to keep the stolen items concealed from potential witness view. The properties of diamonds and jewellery eliminate these issues of concealment and storage, resulting in a very appealing alternative to large bulky items of property.

Virtually Untraceable vs Other Products

The seasoned criminal steals products usually with an advanced knowledge of how he is going to dispose of them. Items like tools, electronics, sports equipment, and other traditionally stolen items are much easier to trace than diamonds and jewellery. Research has shown that pawn shops were far more willing to accept jewellery than electronics because electronics had identifiable features[81]. As an example, a power drill can be described by the company who makes it. That alone can help an investigator search for the stolen product. Also, most products have a model numbers that is the specific type of power drill made by a specific company. In addition, many of the items including tools, electronics, and sports equipment have a serial number unique only to that item that can be used to positively identify the item. Finally, for many people, property they own can be described in these ways but also in reference to marks, scratches, paint splashes, and dents that have occurred on the items and are unique to the specific piece of property. These types of markings are as useful as serial numbers in identifying property and can make disposal of property difficult for criminals.

Pawn shops record the identifying features and even long after the property is sold, if the property is found in someone else's possession, the police will have a good chance of tracking it back to the person who stole it or the fence who sold the property. As a result, criminals take steps to create distance between themselves and the stolen property by using a fence or other means less likely to attract the attention of the police. In doing so, criminals are left to sell the property at a lower price to the fence, because that is the cost of creating the separation between them and the stolen property. The fence now must move the property to someone that the police will likely not uncover. This usually means selling the property privately, or moving it off to police jurisdictions in another province or state. Many people don't record the serial numbers of the property they own. However, those who do, provide this information to the police, who in turn enter that information onto the Canadian Police Information Center data bank (also known as C.P.I.C.). This is helpful because if stolen property is found locally or moves from one jurisdiction to another and then is found, it can be quickly established that the property is stolen by entering the serial numbers of the found property into the C.P.I.C data bank.

Serial numbers are helpful to police, but in some instances can also be a good tool for criminals. The wise fence who takes in stolen property from thieves also has the opportunity to check if the property is listed on the CPIC data bank through use of the CPIC internet web site at www.cpic-cipc.ca. Through this web site, there is the ability to query vehicle serial numbers to find out if they are listed as stolen or not. Vehicles include cars, trucks, snowmobiles, farm equipment, and bicycles etc. This service was made available to the general public, partly so that the public could protect themselves from the purchase of stolen property. If the property queried is stolen and entered on the data bank, the web site advises contact with your local police department. If it is not stolen the web site indicates there are no records found. Serial numbers are one of the greatest tools for police to use to track stolen property, and is evident in the recovery of stolen motor vehicles. This web site can also be used by criminals to check if the property they have on hand is listed as stolen. They can also use it to check on property that they may be about to purchase from a criminal. If the property is not listed as stolen they have a better chance of pawning it off or selling it and not getting caught by police. When a thief brings property to a fence, the fence can check the property on the web site and if it is listed, may tell the thief he will not buy the property and/or to bring jewellery next time. This is not uncommon. Statistics are varied but some show that as many as 80% of criminals surveyed have stolen property to order.[82] That is, a fence asked for a specific item and the criminal goes and steals what the fence wants.

Most diamonds and jewellery cannot be entered on the C.P.I.C data bank because they do not have serial numbers. As for additional identification of the diamonds or jewellery, generally speaking there is no brand name or model numbers that separate one jewellery piece from another. Describing jewellery can be difficult. There are often several hundred or thousands of the same jewellery pieces created by the same manufacturer that look alike. What could be said about a single piece of jewellery that would identify it among those other pieces that look identical? If it cannot be described in some manner that positively identifies it from the other pieces that had been manufactured, then it becomes difficult for police to identify and recover the jewellery. As a result of the untraceable qualities of diamonds and jewellery versus the many traceable qualities of traditionally stolen products, it is easy to assess the desire of criminals to acquire diamonds and jewellery. Recapping the police recovery rates of stolen property helps shed light on this. Although motor vehicles are the number one stolen products by value in the United States, they also have the highest recovery rate, at about 60%. On the other hand, the jewellery and precious metals group with the second highest theft values experience among the lowest recovery rates at approximately 4%[83].

Easy Licit-Illicit-Licit Movement (The Jewellery Cycle)

These same untraceable properties that make diamonds and jewellery desirable to criminals also mean these products can then circulate freely in the jewellery market without detection. This allows criminals to sell the diamonds and jewellery back to the legitimate jewellery market through pawn shops, second hand stores, Internet auctions, auction houses, and jewellery stores. Compared to products that are easily identified, diamonds and jewellery are unquestionably superior in this respect. But the other aspect of diamonds and jewellery is that they are easily morphed into another form, yet still retain their value. A criminal who has acquired a large quantity of jewellery may take to popping stones from the rings and separating them into their categories of diamonds and coloured stones. For large identifiable stones, it has not been beyond the ability of criminals to have the diamond recut so that it cannot be identified. In addition the gold mounts that the stones are popped from are quickly melted down into one large mass of gold. For unsophisticated criminals, they sometimes simply hammer the gold into oblivion so that any inscriptions or engraving on the jewellery cannot be identified. These products, even in their separated state, are just as easy to sell as the complete jewellery piece and to a criminal, just as valuable. What it means is this;

> #1) the ease in ability to first criminally acquire the diamonds and jewellery and then,

> #2) to move the stolen products back into the legitimate jewellery market

The parts that make up jewellery do not decay or break down over time and the components of jewellery are eventually recycled and made into new jewellery. This is what I call the 'Jewellery Cycle'. This concept of jewellery being recycled is nothing new. In fact one diamond expert writes "The truth is, more than half of all marriages fail. That makes for a lot of homeless engagement rings. A lot of people die each year, or go broke, or bankrupt. That's a lot of jewelry for the heirs, the auction block or the pawn shop. Diamonds are forever—but diamond owners aren't."[84] In the 1950's the annual world rough diamond production of diamond was about 30 million carats. This has increased on average at about 1.5 million carats per year to approximately 100 million carats annual production in 2000. Almost all the gem quality diamonds would still exist and are owned by someone or are being resold into the diamond market. It has been suggested by some diamond experts that billions of dollars worth of diamonds are *resold* each year in the U.S. secondary market.[85]

That's right—a recycled diamond. Many people don't really think about this when they buy a piece of diamond jewellery. It is possible that a diamond in a "new" engagement ring actually came from another ring previously. Whether or not a jeweller should disclose this to a perspective buyer is a hot issue in the jewellery industry[86]. In Canada, depending on the interpretation of section 411 of the Criminal Code, disclosing that an item is second-hand may be a requirement for laser marked diamonds and trademarked jewellery that is being resold. When people learn about the recycling of diamonds or jewellery, they can be unnerved in thinking that it may have come from some dead person's estate. Some people even talk about cursed gems—the Hope Diamond for example. Other people don't care about whether or not a diamond is Canadian in origin or if it is a conflict diamond, but if they are determined to get a 'virgin' diamond, brand new, one that has never been on the hand of someone else, then Canadian diamonds are the best choice. This perhaps is one of the greatest selling features of a Canadian diamond and Canadian diamond certificate, because virtually nowhere else in the world can someone make the claim that the diamond they are selling is a virgin diamond. Regardless, recycling is a reality of the jewellery cycle, and the component parts of jewellery do get recycled.

The jewellery cycle is important to understand because what is operating in conjunction with the legitimate jewellery cycle is the criminal element. The following diagram shows the jewellery cycle in its simplest form. The blue circular area to the outside encompasses the dealers and wholesalers of jewellery products, retailers, consumers, pawn shops and auctions. The dealers and wholesalers are where the majority of diamond and jewellery products enter the legitimate cycle. The retailers purchase from the wholesalers and in turn the consumers purchase from the retailers. Eventually, the owner of the jewellery doesn't want it anymore, passes on, needs cash or whatever the case may be, and the jewellery is sold for cash. As mentioned before, a look through the yellow pages directory for Toronto will present numerous advertisements seeking to buy used jewellery.

Used jewellery is sold to the pawn shops, auctions and the retailers. At this point the jewellery may be left intact, especially in the case of estate jewellery, or broken down into its component parts. The pawn shops sell the diamonds back to the wholesalers, dealers, and sometimes jewellery retailers. The precious metals are also recycled but handled differently. This is the jewellery cycle in its simplest form.

Criminals are able to acquire diamond and jewellery at several points on the jewellery cycle. This is evident in reading newspaper articles of jewellery store heists, residential break and enters, and through industry fan outs regarding diamond and jewellery losses. These losses occur primarily through thefts or robberies of jewellery retailers and consumers. This is shown with arrows moving from these acquisition points on the jewellery cycle to the criminal box at the center. This is where both small and large heists occur and criminals can get away with anything from a few hundred dollars of product to potentially millions of dollars of diamond and jewellery. This criminal activity includes credit card fraud, grab and run thefts, residential break and enters, or robberies. The acquisition by criminals can also be through a personal robbery, a home invasion, or residential break and enter. Break and enters are common occurrences in any municipal location and some studies show that diamond and jewellery are among the most sought after items in these occurrences.[87]

Criminals involved in residential break and enters, often trade the stolen jewellery to drug dealers for drugs[88]. The drug dealer then can hold the jewellery as wealth (proceeds of crime), trade it for other drugs or other illegal products, or dispose of it for cash through re-inserting it into the jewellery cycle. Once criminals have acquired diamonds and jewellery from the legitimate jewellery cycle through theft or robbery, there are a number of insertion points into which the criminals can re-introduce the now illicit jewellery back into the jewellery cycle. This is shown on the diagram by the arrows moving from the acquisition points to disposal points. Many of the disposal points are also where the illicit product is re-inserted into the jewellery cycle. Popular points for disposing of stolen jewellery is through fences, pawn shops, and jewellery stores.[89] For instance, criminals can personally sell the diamond jewellery to pawn shops and jewellery stores. This was a popular method used by Bill Mason, infamous master jewel thief, to move the products back into the legitimate jewellery cycle[90]. This is beneficial to both the jewellery store in buying the product off the street, in that they can get nice jewellery or stones a lower cost than what they would pay wholesale. However, the jewellers have no way of knowing if they are buying illicit jewellery from a criminal or licit jewellery from someone who simply wants to sell. In addition, jewellery stores are not required to obtain the name of the seller, like pawn shops are. There is also the potential for criminals to sell the diamonds direct to the dealers and wholesalers. However, there are far fewer diamond wholesalers on the jewellery cycle than pawn shops and retailers, and this end of the business is somewhat more insulated from the majority of criminal elements than other points on the jewellery cycle. What is more common is that the pawn shops act as unwitting or deliberate collection points for stolen diamonds and in turn sell them off to diamond dealers and wholesalers.

Our Disposable Society

We live in a disposable society. Cell phones, electronics, computers and techno gadgets are outdated at barely 6 months old in what is called planned obsolescence. The new gadgets coming into the market are known as "hot products" and are highly sought after by criminals and usually remain so until the next model emerges or they otherwise loose their appeal.[91] Today it is cheaper to purchase a new replacement models than to have old electronics updated or repaired. Society was not always like this. A few decades ago if the stereo blew a transistor, it would be fixed at the local repair shop. Going back a little further, it was not uncommon for a television or radio to blow a tube. Consumers purchased a new tube and installed it themselves. As a society we now have access to more electronic gadgets than ever before. With advancements in productivity, foreign wages, and low costs to produce these gadgets, repairing these items in North America is prohibitive because labour costs here are so much higher. It then becomes cheaper to just buy a new product rather than repair the old one, so people simply dispose of the old one. No one bothers to repair the old VCR's. Instead they buy a DVD player. The same goes for the cassette player that replaced the 8 track player, that got replaced with a CD player. Electronics are a product that criminals have been eager to acquire for several decades.[92] However, with electronics becoming somewhat disposable and quickly dated, there is a diminishing value for products that are second-hand or otherwise used. This is reflected in a lower value obtained in selling the products and increasing difficulty in selling dated electronics.

A traditional venue for criminals to sell such items is pawn shops or second-hand stores. The pawn shop takes a risk in buying these items because the longer they are on the shelf, the less money they receive when and if they do sell the item. This shift in how dated electronics are dealt with has been evolving over the past twenty years. In the last 10 years, it has become quite apparent. With this, the ease that criminals were once able to sell these dated electronic has gone. Alternatively, the ease with which a criminal can sell diamonds and jewellery has not been affected by these downward pressures, and the pawn shops and second-hand stores may be more likely to purchase diamonds and jewellery now than in years gone by to fill the electronics gap.

Foot ball field size mountain of used Computers and Televisions. The barricade in front is three feet high.

The Big Screen TV

For the street level criminal doing break and enters into residences, the television has been a staple on the plate of acquisitions[93]. Nearly every home has a television. They are valuable and typically easy to sell. Virtually every pawn shop has one, and usually more, for sale. Unfortunately for criminals, the television has evolved over the past twenty years and has resulted in an item whose acquisition is less desirable than ever. In the seventies, the colour television was a hot commodity and a very expensive household item. Many people were still watching a black and white TV and some didn't have televisions at all. This carried on into the eighties and the next advancement was the remote control. First the remote control was connected by a wire to the TV, and then it became wireless. These advancements continued to add value to the television and the relative price was high compared to other household electronics. Criminal acquisition of televisions continued to mean good money in their pockets. However, this began to change in the decade following 1990. The same advancements in technology, productivity, and foreign imports that made repairing electronics futile have also served to lower the price of televisions.

At the beginning of the 1990's, a 29 inch television was considered a big TV and it was priced at about $1,000. A criminal who stole one of these could likely expect a 10% return on the sale of the TV to a pawn shop, which would bring him about $100. As the years continued on, the price of these TV's didn't go any higher. In fact they began to decrease to where now a new 29 inch TV can be purchased for less than $300. Given a 10% return on the sale of the TV, the criminal profit has gone down significantly. Also, taking inflation into account between 1990 and 2004, the value a criminal could get for a new 29 inch TV has eroded even further.

A 29 inch TV is not an easy item to carry, and manufacturers are making TV's a lot bigger today. It's not uncommon for a residence to have a television that is 40 inches or bigger, and these are nearly impossible for one person to carry. It requires quite an effort for two people and something bigger than a compact car to transport. The declining values for televisions translates into a diminishing appeal to criminals. As well, the increasing sizes of televisions means they are just that much more difficult to steal than ever before. To make this point, when compared to a low-end piece of jewellery that will easily fetch $30 dollars at a pawn shop or traded for drugs, diamonds and jewellery presents an appealing alternative to televisions that have historically been a priority on the criminal's shopping list.

The High Value Returns of Diamonds and Jewellery

The last point discusses the value of diamonds and jewellery. For the criminal committing a theft, the higher value of products they steal should result in a better return of cash or trade. They also want to get the maximum dollar value for each item. In any household, there are typically very valuable tools, appliances, electronics, and even cash. Of the items that are typically stolen, the refrigerator and stove usually are not part of the picture. This leaves stereos, televisions, tools, electronic games, telephones, microwave ovens, and similar items. This is certainly the case for electronics. A criminal pawning a $2,000 dollar stereo that is in brand new condition can expect to obtain about $200 for it. Given that a lower-end, 0.50ct diamond ring or pendant could easily have a retail price of $2,000, using the typical pawn shop pay out for products of 10%, the ring should fetch about $200. However, generally speaking, this 10% rule holds true for nearly ever item in a pawnshop except diamonds or precious metals. This higher return for jewellery vs. other products is reflected in studies conducted into the criminal use of these products.[94]

Diamonds represent an opportunity for the pawn shops to make some quick money on an item that has been pawned and as a result, some are willing to pay more than the usual 10% rate to obtain these goods. The larger pawn shops and pawn shops that operate as a chain will often base the amount they are willing to pay for the diamond on a rising linear scale based on the increasing size (carat weight) of a diamond. I have seen on many occasions that when they take a diamond in on pawn, they measure the diamonds diameter to establish its carat weight. For instance a round diamond of 4mm diameter is approximately a 0.25ct (25 point) stone, a 4.4mm diameter is about a 0.33ct (33 point) stone, a 4.8mm diameter is about a 0.40ct (40 point) stone and a 5.0mm diameter is about a 0.50ct (50 point) stone. The clerk at the pawn shop will measure the stone with calipers to establish the carat weight, then match the established carat weight up to their chart that tells them what they will pay for the diamond. The chart that many pawn shops use to decide how much to offer for a diamond goes something like this. For diamonds that are 0.25ct–0.33ct, they offer $4 per point, 0.33ct–0.40ct they offer $6 per point, 0.40ct–0.49ct they offer $8 per point, and for anything 0.50ct or greater, they typically offer $10 per point. Using this chart, the 0.50ct (50 point) diamond ring that normally retails for about $2,000 would fetch about $500 (50 points x $10) at the pawn shop. This reflects a 25% return on the value of the ring vs the usual 10% return that most other products would claim. Generally speaking, the pawnshops don't focus much attention on the colour or clarity in fixing a price; they simply use the weight of the stone. Diamonds and diamond based jewellery represents greater rates of return for their values than do other items that have been historically stolen.

Nearly Everyone Has Jewellery

The North American diamond culture means that almost everyone has jewellery. It is estimated that 78% of Canadian women own at least one diamond[95]. However, the personal acquisition of jewellery goes beyond the desire for diamond jewellery and marketing, to the customs and traditions of society. The one tradition that puts gold on the finger of nearly everyone is the ring exchange between men and women when they get married. This act of marriage and the diamond engagement ring that precedes the wedding rings means that over 70% of married couples own some precious metals and diamond.[96] These traditions and customs lay the foundation for good marketing strategies that promote gifts of jewellery on anniversaries, mother's and father's day, and birthdays. This is also carried on to children with the parental purchase of jewellery for everything from baptisms to a 16th birthday to graduation gifts. Some statistics show that 30% of teen diamond jewellery is obtained at birthdays and 48% is obtained at Christmas.[97] The adults that have come together in marriage usually bring with them a number of jewellery pieces that they accumulated since birth. This jewellery is then added with those pieces accumulated through marriage and subsequent anniversaries and birthdays, creating a wealth of jewellery in each household.

Depending on the income of the household, the cumulative values will vary greatly from house to house but regardless, they will almost always have jewellery.

For the criminal stealing jewellery, there is no hit and miss with a residential break and enter. If criminals come away from the residential break and enter without any jewellery, it is probably because the owners took the time to hide it well. If you are at home right now, do yourself a favour. Go to your bedroom and take the jewellery off the night table or dresser drawers and find a place to hide them, preferably not in your room. If your house ever gets robbed one of the first things a criminal is going to look for is jewellery and they will look for it in your bedroom[98]. This abundance of jewellery in every house in every city and town in North America means that once a criminal has established modes of selling the jewellery, they virtually have an unending supply.

In terms of personal robberies, the same holds true. A criminal robbing someone at gunpoint could steal the wallet, watch and rings. A typical man's gold wedding ring without a diamond would fetch the criminal about $25–$30, the watch $30 or less, and whatever cash. Most robbers will likely just take the cash and jewellery and dump the rest of the wallets contents. Hanging on to these items is a sure way to get caught with identifiable property. Unless there is a significant amount of cash obtained through the robbery, the main value is in the jewellery.

Cashless Society

I normally carry about $15–$20 cash. That's it. There was a time when I use to regularly carry over a hundred dollars and by some standards this is nothing, but those were pre-bank machine, instant teller, and debit card days. In fact, Canadians are the worlds highest per capita users of interact, cash machines and debit card transactions.[99] We are increasingly moving towards a cashless society[100]. For anyone living in an urban center in North America, the need to carry money for day to day activities has diminished. It is much simpler to get cash as needed from cash machines, or to use a debit card for purchases. There is also the lower risk of losing large amounts of money. For a criminal looking for cash in a personal robbery, home invasion, or residential break and enter, the days when large sums of money were easily obtained are gone. Electronics and televisions are losing their value and in the case of televisions, are increasingly more difficult to steal. The diminishing values are also seen in electronic games, appliances, and tools. Criminals now need to focus on valuable items that they can readily get their hands on, are difficult to trace, easily hidden, hold their value and are easily sold, and this means diamonds.

Although I have included this issue under the heading of micro issues, this issue is just as important as a macro issue. In carrying less cash than ever, in a reciprocating fashion, people make that many more transactions through the electronic formats. Likewise it is far more common now to make large cash transactions via an electronic funds transfer or related electronic transaction. These electronic transactions are all trackable and as such are problematic for criminals that are trying to conduct business under the authorities' radar. Financial transaction recording legislation and further increases in electronic transactions are hurdles that criminals will have to overcome. Criminal use of alternate currencies or conducting financial transactions through the use of diamonds and like commodities is not only appealing but may become increasingly necessary to conduct business.

Robbing a Jewellery Store

Much of what has been focused on discussed street level crime that is committed through residential break and enters, home invasions, and personal robberies. The other locations that diamonds and jewellery are also readily available are in jewellery stores, wholesalers, and traveling jewellery persons. However, like a home invasion or personal robbery, robbing these intermediate levels of the diamond industry takes a more brazen and organized criminal. Because of the employees and customers there is an element of danger in these crimes beyond that of strictly a property crime. Jewellery stores, wholesalers, or traveling jewellers will often carry several hundred thousand dollars in stock, and as a result they become a great target for the criminal that is going for a big score. In a robbery or attack on a jewellery store, it is not uncommon for criminals to come away with over $100,000 worth of diamond jewellery.

I spoke with a criminal who made a career of robbing jewellery stores and asked him "Why jewellery stores?" He said "It is easier to knock over a jewellery store than a bank". He said the security at jewellery stores was typically low, the expected haul was high, there was no marked money, or exploding die packs typically involved in robbing a bank. He also said that by the time the jewellery was sold he could easily come out on top financially over what he would get for robbing a bank.

6

Criminal Use of Diamond

There is so much potential for the criminal use of diamonds that the opportunities are only limited by one's imagination. The criminal activity described in this chapter and the Chapter: "Treatments, Scams and Misrepresentations" could be conceptual or actual. For the most part, these activities could be conducted through the use of rough or finished diamonds but in reality the vast majority of diamond specific criminal activity in Canada is committed through use of finished diamonds. This stands to reason because rough diamonds have only been mined in Canada since 1998 and as a result, this part of the industry has not evolved enough to present the opportunities that the finished diamond market presently affords. For one thing, there is not that many opportunities for criminals to acquire rough diamonds in Canada yet. Secondly even if criminal was able to acquire rough diamonds they have limited opportunity to use them. On the other hand, finished diamonds have been bought and sold in Canada since the country's creation in 1867. In fact, the Canadian jewellery industry can be illustrated by the publication, The Canadian Jeweller, which is over 125 years old and Canada's longest running magazine. The same properties that make diamonds valuable today also existed back then and as such, diamonds and jewellery have been historically targets of criminal acquisitions. One of the primary factors limiting the exploit potential of rough diamonds in Canada is the limited black market for rough diamonds. Referring to the jewellery cycle, there is little opportunity to insert stolen rough diamonds into the legitimate diamond and jewellery market. This is because the rough diamond market exists further up the mine to market chain than where the jewellery cycle exists. Finished diamonds have few limitations of this sort and as such there is prolific criminal activity within this market sector. For many of the crimes, the small compact value of diamond is the exploitable factor. For some other crimes it is the valuation; for others it's the untraceable qualities; and again for other crimes, it's the lack of knowledge that most people have about diamond that criminals can exploit.

Whatever the case may be, criminals will seek out and exploit the weakness in any system.

Diamond Cutting

There are few diamond cutting enterprises in Canada but many in North America and these are for the most part, well-run operations. One of the drawbacks to cutting diamonds in Canada is the high cost of labour and the limited labour pool with skills to cut diamonds. To help eliminate this problem, diamond cutting factories in Yellowknife have brought in experienced diamond cutters and polishers from foreign countries where the wage for cutting diamonds is a fraction of what is being paid in Canada. So not only do Canadian diamond cutting factories have a labour squeeze, they also have a profit margins squeeze due to the higher labour cost of cutting a diamond in Canada. This can translate into an extra $200 per carat to produce finished diamonds in Canada versus let's say India, where the cost may be less than $30/carat for example. The way Canadian diamond cutting factories get around this is through cutting only high-end diamonds and government assistance. The larger and higher quality the diamond the better because the high cost of diamond cutting in Canada can more easily be incorporated into production price. Generally speaking it is only higher quality diamonds larger than 0.25 carat that are produced. With the small margins and labour squeeze, Canadian diamond cutters are already behind the eight ball. Adding a fluctuating Canadian dollar, a rough diamond shortage to the cutting factory, and the introduction of a criminal element to the mix, and the combination of events could provide greater potential to spur on criminal activity. This sector of the mine to market chain is exploitable as is every link in the chain, but like the mining sector there have been few recorded criminal occurrences. It appears that at present, this end of the diamond industry in Canada is somewhat insulated from criminal activity.

At this stage of the mine to market process, any criminal involvement within the diamond industry would be difficult for law enforcement. This is because at the cutting stage of diamond processing, the rough diamond is changed into a finished diamond. There are several aspects to this process that are open to exploitation not the least of them being the grade of diamond. Rough diamonds are graded as best as possible based on what a grader can see of the stone's characteristics. However, there are no master stones to capture exactly what the colour of the rough diamond is and perhaps even more difficult is gauging what the clarity is of the diamond. Many rough diamonds are beautiful crystals that a grader can easily look into and establish what the clarity would be of the stone. As previously mentioned, there are rough diamonds with skins or a coating that cover them. The skin or coating is a natural feature on some diamonds and the grader can only see into the stone once part of the coating has been polished away. This is what is known as polishing a window. However, even with a window polished into the stone the diamond grader can see into the stone but not necessarily through the stone. The stone may be nearly inclusion free on the inside yet there is still the skin on the other side that impedes a good look at the inclusions. Finished diamonds may be easily over-graded or under-graded to the benefit of a criminal enterprise. However, the opportunity for a criminal operation to over-grade or under-grade a rough diamond provides equal exploit potential. For a large criminal operation the over grading of stones allows the importer to claim a higher cost on the stones and if their objective is money laundering then this works well for them.[101] Under-grading the diamond, rough or finished, allows for a lower valuation of the diamond on import and a lower tax bill on the imported products.[102] These diamonds do not even have to be falsely graded for criminals to have the valuations of diamond work for them. Instead of falsely grading the diamonds the importer can strike a deal with the exporter of the diamonds and simply have the exporter write whatever value per carat they want for the diamond parcel on the invoice. There are few people in North America that can put a value on rough diamonds so in turn few people could refute what the invoice accompanying the parcel says. There are other potential criminal activities that can be engaged in at this stage.

When a diamond is cut there is a huge loss in weight from the stone that cannot be physically accounted for. This part of the diamond that is essentially lost provides a great opportunity for substitution of stones. Switching low value stones for high value stones one method of criminal activity[103]. This can happen at various stages in the processing of diamond. If a criminal wanted to switch low-end diamonds from one country with high-end diamonds from another country this would be one opportunity to do it.

As of an example of what can happen,

A diamond cutter starts with a nice diamond crystal that weights 1.00 carat. It is estimated that this diamond will have a 50% recovery weight and as such they'd get approximately 0.50 carats of finished diamond. Typically this recovery weight would result in one large stone and a small stone or two medium sized diamonds of equal size. To produce the finished diamonds, the stone is first cut with a laser or diamond saw or perhaps it is even cleaved to remove the bulky unwanted parts off the diamond. For simplicity sake, this bulky unwanted part of the rough diamond can be called the 'unwanted rough diamond'. Cleaving a diamond is like taking a hammer and chisel and chipping off a part of the diamond. This step is similar to what a stone carver does at the beginning of carving a statue to remove the bulky unwanted part of the stone. Whatever process is used, it will take a good portion of the bulky outer material off the rough diamond but the rest needs to be polished off. Through sawing and cleaving the 1.00 carat rough diamond, 0.30 carat weight of the rough stone is removed. These pieces could be kept or be thrown away if they were not worth processing further. After that, about 0.20 carat of the rough material would be polished away, to end up with 0.50 carat of finished diamonds. The polishing process that creates the diamond facets eliminates 0.20 carat worth of rough diamond material. This 0.20 carat weight of diamond material is polished away and it can be called diamond 'lost to the polishing wheel'. This diamond material that is lost to the polishing wheel, is usually flushed down the drain. In a perfect world, it would be possible to perfectly reconcile the rough diamond weight with the finished diamond weight. However, because of the material lost to the polishing wheel it is not. The cutting and polishing process of turning a rough diamond into a finished (cut and polished) diamond is considered a weak link in the chain of custody of diamonds.[104]

1 ct rough diamond = 0.50ct finished diamond +
0.30ct unwanted rough diamond +
0.20ct lost to the polishing wheel

Once the stones have been cut there is no information as to what country the diamond came from or what the original weight was. The fact that a portion of the rough stone is lost to the cutting wheel means that any unscrupulous diamond cutter could claim whatever amount he wants as being an amount lost to the cutting wheel. As such the criminal operation could reconcile the finished diamond and the unwanted rough diamond material with the weight of any finished diamond. For instance, if the 1.00ct diamond described above was very low quality and the criminal wanted to substitute that stone with another diamond that was a top quality stone of 1.05 carat, he could show finished diamond that weights 0.55 carat. He may actually have 0.25 carat of unwanted rough material and could claim that 0.20 carat was lost to the polishing wheel. The weight adds up to 1 carat and no one can argue the amount lost to the polishing wheel. Providing the criminal doesn't claim too large of a finished diamond relative to the rough diamond weight it would be difficult to uncover this. Criminals could actually cut rough diamonds of foreign origin and substitute them for whatever he has in stock. The benefit would be that the goods he produced are higher end diamonds vs the lower end stones he had in stock or what he reported as having in stock. The low-end diamonds could be cut and polished, traded, and sold as rough or finished products. The criminal simply has to sell the original diamonds or use them in other schemes and the loose ends are sewn up. On the other hand if the cutting facility is run by criminals and cuts diamonds on demand for clients, other criminal activities can be employed. If a client supplies rough diamonds of high quality, the stones could be cut and polished and the high quality gems substituted with low quality stones that are then sent back to the client.

Although it is easy to employ such a fraud the question is why would someone wish to do so. Diamond substitution can be a function of several criminal motives including tax evasion, the result of higher profit margins with, and demand for, higher end goods, or a means to launder illicit goods or non-Kimberley Process goods.[105] Substitution of stones is one method of criminal activity, however the diamond cutting process simply makes it more difficult to uncover criminal activity through substitution.

In the past, the industry could only estimate the recovery weight of a diamond. Using the example of the 1.00ct diamond with the expected recovery weight of 50%, in reality once it is finished it may only have a 45% recovery weight or maybe the recovery weight is higher and turns out to be 51%. The final weight depends on many factors including the stone shape, how many finished diamonds will come from the single rough diamond, the internal crystal structure, inclusions, what plans the diamond designer has for the stone, and the cutting and polishing process. Today the process is a little more exact due to computer technology that allows for external mapping of the rough diamond. By mapping the exterior of the rough diamond the computer is then able to predict the optimum cutting shape of the rough diamond to obtain the highest yield of diamond. With this technology the finished weight of the diamond matches what it is expected to be 80–90% of the time.[106] What this means is that through this technology one can say with some reliability what the final recovery weight of the diamond will be. This technology helps in assessing the recovery weight and is widely used by the diamond industry and gaining further acceptance, yet it is not a requirement.

Regardless of this new diamond mapping technology there is enormous potential for other criminal activity here. Potential includes:

    a.   Inserting illicit finished diamonds back into the legitimate market stream as if they were diamonds cut at the factory. The illicit diamonds are simply added to the inventory.

b.   Smuggling and inserting conflict diamonds into the legitimate diamond market.

c.   Manipulating expected recovery weights to allow for the insertion of illicit rough diamonds into the inventory. For example, a diamond-cutting factory has on inventory a 200 carat parcel of rough that is only expected to yield a 25% recovery weight or 50 carats of finished diamonds (200ct x 25% recovery weight = 50ct). If the factory falsely claims this 200 ct parcel on inventory has a 50% recovery or 100 carats finished, then the factory could push through a second smaller parcel. Perhaps this second parcel is a smuggled parcel of 100 carats of rough diamonds with a recovery weight of about 50% or 50 carats. These finished diamonds can then be added to the original first parcel and claim a total recovery weight of 100 carats or 50% for the original first parcel.

Criminal activity of this nature would be very difficult for law enforcement to uncover and deal with in. In Canada's relatively unregulated and uncontrolled diamond industry, this would be even more problematic.

The cutting of diamond can also be used to bypass the Kimberley Process (KP). This could be done through cutting the rough diamonds at the source. If diamonds are mined in a country and then immediately cut in that country, the finished diamonds don't require KP certificates. The KP certificates are only required for rough diamond export or import. Once the diamonds have been cut and polished they can be sent just about anywhere in the world. If criminals cut diamonds in the Congo or Sierra Leone, the heart of blood diamond country, not only is there an unlimited supply of illicit and or conflict rough diamonds, the KP process is bypassed. And the diamonds don't really have to be fully cut and polished. All that is required is some basic cuts, otherwise known as 'blocking' the stone, that prepares them for the next step of cutting and polishing. Once a few basic cuts have been made to the original rough diamond, it is no longer considered a rough diamond and no longer subject to the KP regulations. Regardless of the blood diamond issue, cutting rough diamonds at the source country becomes even more profitable in some countries than cutting diamonds elsewhere. This is because there is often an export tax on rough diamonds. Cutting diamonds in the source country eliminates this export tax on rough diamonds and the need to follow the KP rules. Organized crime could easily set up cutting facilities in jurisdictions where some authorities are corrupt and willing to allow criminal opportunists to exist. These types of cutting operations could easily support illicit mining operations run by criminals and rebel groups or smuggling operations from other jurisdictions or neighboring countries.

Rough diamonds have a relatively limited market for resale in Canada and North America. It is certainly a small fraction of the finished diamond resale market. But as previously shown, rough diamonds are apparently illegally entering Canada via courier services. Criminals have also attempted to smuggle rough diamonds into Canada and while some have been caught, law enforcement cannot catch all the rough diamond smugglers. Despite the potential for criminal activity in the diamond-cutting sector, at present this section of the Canadian industry is among the least likely of sectors to be criminally active. There are many other ways smaller parcels or individual rough diamonds can be utilized. If a jeweller is presented with rough diamonds from someone off the street, there is great potential to use the rough diamonds. There is nothing illegal with buying rough diamonds from someone. As a business venture, if the product sold by the vendor can be obtained at a lower price it is a fruitful venture. The jeweller would likely want to have the stones cut so he would have a better chance of selling them, however some jewellers have taken to incorporating the rough diamonds into jewellery. Depending on the quality of the stones and the price they can get them for, it is actually quite easy to have rough diamonds cut in Canada or in the United States. Virtually any Canadian jeweller could have a 2.25ct rough diamond cut for about $190 U.S. ($85/carat) or about $215 Canadian. In the U.S. there are many reputable cutting and polishing facilities that will do this, and there are some in Canada. At a 50% recovery rate, this would leave them with a 1.12ct of finished diamond. For simplicity sake, say the finished diamond weight was made up of two (2) diamonds. They are 1 x 0.85ct and 1 x 0.27ct round brilliant cut diamonds for a total recovery weight of 1.12cts.

What could a jeweller do with a rough diamond? If the jeweller purchases a nice rough diamond off the street from a criminal, the criminal may have smuggled the rough diamond or stolen it, but the jeweller doesn't know this and has no way to determine this. Even if the jeweller asked a few questions the criminals would likely lie about the origin of the stone. If the rough diamond is a lower quality stone of about 2.25 carats in weight in the SI clarity and H colour range, perhaps he pays about $1125 (2.25 carats x $500/ct) for the stone. This could be an expected price for rough diamond that is of illicit origin. The jeweller can have the diamond cut as a round brilliant for another $191.25 (2.25carat's x $85/carat). After cutting, they are left with two (2) finished round brilliant cut diamonds. One is 0.85 carat the other is 0.27 carat and both happen to be SI2 clarity, H color. The total cost of these stones including the cutting costs is about $1,316.25 U.S. ($1,125 for the rough diamond and $191.25 for the cutting cost). Buying these stones from a diamond dealer, the jeweller would normally pay about $1800 U.S. for the 0.85ct diamond and about $250 U.S for the 0.27ct diamond before taxes for a total of $2050 U.S. for these two diamonds. This translates into $733.75 U.S. worth of savings on the two stones. Similarly, a criminal who illegally imports or smuggles the stones into Canada could have the stones cut and sell them as finished diamonds or sell the rough diamonds to another criminal who could do the same. Opportunities abound.

Diamond Switching or Mixing

Diamond switching is a way to insert illicit rough or conflict diamonds into the legitimate diamond market.[107] Mixing of rough parcels before being imported into a country is also recognized as a vulnerability in the diamond markets.[108] When diamonds are exported from Canada or any other country signed onto the Kimberley Process, an application for a Kimberley process certificate must be obtained. You can find Canadian Kimberley Process certificate and application form information at www.NRCAN.gc.ca/kimberleyprocess. The criminal operating with legitimate and illicit rough diamonds could complete the application stating that the parcel he is about to export has for example: 100 carats of rough diamond with a total value of $50,000 U.S. or $500/carat. The criminal shows documentation of imports or other legal acquisition for these stones and receives the required certificates. The parcel is prepared for shipment, however the actual diamonds that are shipped are not of the stated value as reported to obtain the Kimberley Process certificate. The parcel is exported to, for example, Venezuela, a rough diamond producing country where the criminal's partner operates illegally in the diamond industry.

As a sidebar, Venezuela is used only for example sake, but it is noteworthy that diamond smuggling and illegal diamond mining is rampant in this country[109]. Illegal diamond mining is another opportunity for criminals to profit from diamonds[110] and some countries are beginning to crack down on illegal diamond mining[111]. Criminally and geographically speaking, Venezuela is interestingly positioned between Guyana (a former Dutch colony) to the east and Columbia, the world's largest cocaine producer, to the west. Officially, Venezuela has exported no rough diamonds since January 2005 and the implementation of the Kimberley Process, which begs the question, where have the diamonds gone?[112] Yet over the last decade, production of diamonds in Venezuela has diminished from over 300,000 cts/year to 30,000 cts/year, the neighboring Guyana has seen its diamond production increase from 40,000 cts/year to 400,000 cts/year in the same time period[113]. In reality the official production in Venezuela appears to have gone down but unofficially it has remained the same, however, the diamonds are no longer appearing as production for Venezuela as they are being smuggled into Guyana.

As the scenario progresses, the parcel is received by the criminal in Venezuela and the entire parcel of diamonds is switched or several of the poor diamonds are switched with high end diamonds of equal weight. The parcel is then sent back to Canada as returned goods. All the Venezuela partner has to be sure of is that the weight in carats is 100 carats so that this corresponds to what the certificate says. This can be done by simply adding a few small diamonds to the switched parcel until the weight is the same. When the parcel gets back to Canada it is checked by the Canadian Border Services Agency, the weight is the same, and matches corresponding documents. In addition the values are the same because the diamonds that are switched match what the price per carat indicate on the certificates.

In this scenario the criminal activity would normally get stopped when the parcel enters Venezuela. This is because Venezuelan Customs officials should examine the parcel. They would check that it has the Kimberley process certificates, which it would have from Canada. They could weigh the parcel and would see the weight is 100 carats as the parcel says. They could even test the stones to be sure they are diamond. However, I suspect that few if any Venezuelan Customs officers would be able to make the assertion that the value stated on the certificates for the parcel is anything close to the true value of the parcel. Very few people in the world, let alone customs officials in most countries can determine the value of a bulk rough diamond parcel. Even in Israel, where all parcels are inspected, government valuators find it difficult to be precise in valuation.[114] In the end, if the parcel has the KP certificates, it weights 100 carats or what ever the stated weight is on the KP certificate and they are diamonds, then the parcel will undoubtedly go through without a snag. The switch is made in Venezuela and the parcel is sent back to Canada as returned goods.

Now the diamond parcel is back in Canada with the upgraded more valuable diamonds. In this case the diamonds are illicit and have been illegally inserted into the legitimate rough diamond market. The criminal has moved hundreds of thousands of dollars across borders and skirted import taxes on these items. The diamonds could just as easily be blood diamonds inserted into the rough diamond market that now have skirted the Kimberley Process. It is important to make that distinction here. Just because the rough diamonds don't have Kimberley Process certificates, doesn't mean they are blood diamonds or conflict diamonds in respect of the spirit of the Kimberley Process. I would suggest that a parcel of Canadian mined rough diamonds stolen from Canada but traded criminally without KP certificates and then inserted into the diamond market in another country could not rationally be considered blood diamonds. Illicit diamonds yes, but not blood diamonds.

The entire Kimberley Process and its evolution brings up some interesting questions. If one cannot with any scientific certainty or definitive conclusion determine the source country of a diamond, with out personally following a stone from mine to market, how can a diamond be authenticated with absolute certainty that it is conflict free? The answer is that we don't know for sure. This is exactly why the research being conducted by the RCMP forensic lab on diamond profiling is so important. This research is aimed at being able to determine with scientific certainty what country a particular diamond parcel comes from.

Sales

To be perfectly clear at this point, criminals in the jewellery industry are but a fractional representation of the whole. This criminal involvement is no different than what is found in other industries. However, criminal involvement at the retail and wholesale level of the diamond industry provides its own opportunities. One of the greatest opportunities is the ability to retail the stones without the 6% Goods and Services Tax (GST). It is quite easy for a criminal to do when operating as a retail jeweller. He simply has to acquire diamonds off the street, through pawn shops, the Internet, auctions, off account from wholesalers, or a host of other avenues, and then offer them up for sale in his shop. These diamonds purchased would not be entered onto inventory and if bought off the street, auctions, or through pawn shop contacts and they would have typically been acquired at ridiculously low prices, well below wholesale pricing. The diamonds can be retailed in the shop through cash sales and not only does the criminal reap huge profits on the turnaround of the stone but again there is no GST on it or income tax payable as it goes unclaimed and off the ledger books so to speak. This is also valuable to the money launderer who wants to purchase diamonds without the purchase being reported.

A criminal could also sell fakes (a.k.a simulants), stolen diamonds, or manmade diamond synthetics as genuine diamonds to unsuspecting buyers and make profits on these frauds. I once went into jewellery store in Calgary and noticed a diamond that was priced at a very low price for the quality of stone. I asked the clerk to see the ring and when I got it close, I knew immediately it wasn't genuine. I had a second look through their store loupe to confirm my belief and told the clerk that the stone wasn't a diamond. The clerk became very indignant, as one would expect, that someone was calling this stone a fake. I told her if she didn't believe me to go get her thermal tester, and check it out.

A thermal tester is an instrument that can help tell the difference between diamond and diamond fakes. It relies on the ability of diamond to conduct heat better than any other gemstone. When the instrument is turned on, a small metal tip is heated up and then the tip is placed on the diamond. The heat conducting ability of the diamond draws the heat away from the tip of the instrument at a rate that is calibrated for diamond. When the instrument registers the heat transfer coincident with diamond it usually beeps or a light flashes or both depending on the type of instrument.

The clerk used the thermal tester and the stone did not test positive for diamond. She was shocked and apologetic and I truly believe that she didn't know. However, someone must have known. Was it the store owner, the stone setter, the gemologist? Was this ring intended to defraud someone? Regardless, someone would have gotten ripped off in purchasing that ring. Further to this, a store could simply be a front for a criminal organization and specialize in moving stolen product from the black market into the legitimate market or defrauding people.

In the last chapter, discussion surrounded the acquisition of diamonds and jewellery through thefts, break and enters, and robberies. The question is where does this jewellery go, especially the diamonds that are laser marked with serial numbers? Many people have concerns about the theft or loss of their diamond, especially those who have high value stones. Many see the branding and application of serial numbers as an answer to this. One of the greatest selling tools to come on the diamond market was the advancement of laser technology that would allow for the inscription of a serial number and brand logo on a diamond's girdle. The laser branding is done on the girdle so it does not affect the colour or clarity grade of the stone.

The branding of the diamonds means that a potential buyer may make return trips to the jewellery store because it's not just any diamond; it was a diamond from that particular store. As important to branding the logo on the diamond, is the serial number that most brands also inscribe on the stone. The serial number corresponds to the company records kept regarding the stone and also gives a degree of security to the stone. If the stone is lost or stolen then the serial number can be used as positive identification of the stone and for the case of Canadian diamonds, used for authentication purposes and to track the diamond from mine to market. The process of laser marking diamonds involves the focus of laser energy onto the girdle of a diamond to create a pattern in terms of a trademark and for inscribing a serial number. The energy of the laser is sufficient to change the diamond state of carbon into basic carbon. In the basic carbon state the inscription or logo shows up as black carbon on the diamond's surface. Diamond is very hard and would require direct diamond wheel polishing or diamond cutting to remove surface material. However, because the inscription that is made on the girdle of the diamond is now regular carbon and does not possess the same resilient properties as diamond, the inscription can be easily polished off with a high-speed buffing wheel and jeweller's rouge. Jeweller's rouge is a wax-like polishing compound used to buff the surface of precious metal to produce the smooth finish and bright shine. In the simple process of mounting the diamond, the diamond is mounted into a ring and the prongs tightened down on the stone. Then the ring is buffed to smooth the prongs and brighten the finish, but the buffing process can take off the inscription. Cleaning the diamond in a steam cleaner or an acid bath can also have the same effect. In this case the diamond has to be sent back to the supplier so he can verify the diamond, a process he does through checking the proportions and characteristics of the stone and cross-referencing it with what is on record. Once verified, the inscription is then re-lasered onto the girdle. Depending on the supplier, some markings are lasered deeper into the stone and are more difficult to remove. Others suggest that even after buffing or removal of the blackened surface inscription, the markings can be seen through back lighting the stone. Regardless, any evidence of previous marking could be very difficult to detect.

In this regard a criminal who has removed the marking could quite easily sell the stone to a jeweller who is not looking for this or to a jewellery store run by criminals. Even if the marking was present, a criminal could still purchase the stone and mount the diamond in a style of mounting known as a bezel mount (pictured below), where the gold envelopes the girdle and therefore hides the markings. Still, despite the potential downfall, the markings are at least one method of tracking and identifying a diamond.

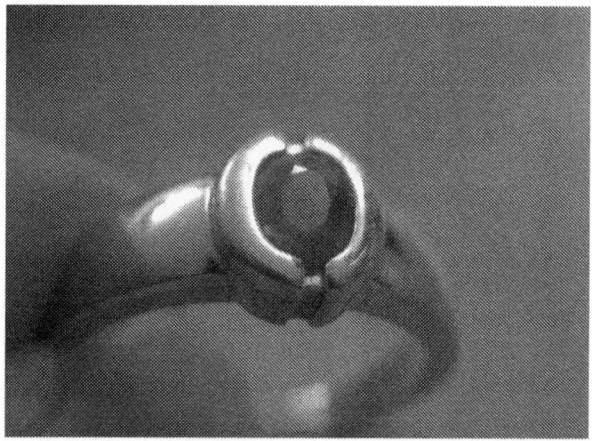

Gemstone in a bezel mounting

Diamond Salting

With reference to criminal activity, this is not unlike the cocaine dealer that cuts the cocaine with cornstarch or baby powder to increase his profits except that there are a few different ways that diamond salting can take place. Some believe that salting diamond parcels is a common practice by criminals that operate in the diamond industry. A diamond jewellery manufacturer that specializes in making 1.00 carat diamond jewellery pieces studded with tiny 0.01ct diamonds purchases copious quantities of these tiny diamonds. Even in a relatively small 10 carat parcel of diamonds that is made up of 0.01 carat round brilliant cut diamonds there would be an enormous number of stones. In fact there would be about 1,000 tiny 0.01ct diamonds in this parcel. If each diamond was tested simply on a thermal tester to see if it was genuine, this would take about 1 minute for each stone. To test the entire parcel would take about 16 hours of work. This time allotment would significantly impact the cost of the diamonds and as a result, the diamonds of this size are rarely tested. What may be done is a spot test on sample of the parcel, but this is a hit and miss method. Criminals understand this and can add a quantity of cubic zirconium (CZ) to the parcel and commit fraud. The CZ's are then unwittingly added to the diamond jewellery by the manufacturer and sold to the public. Similarly, rough diamond parcels can be salted with non-diamond rough minerals. It is not uncommon for rough diamond buyers that are unaware of the properties of diamond or inexperienced at dealing with rough to get taken by criminals who will knowingly sell them a fake.

One of the best examples of rough diamond fakes that I have ever seen are pictured below. These fakes are CZ's that were cut to resemble the classic octahedral crystal shape of a rough diamond. The criminals even took to cutting surface features into the fake stones that mimicked those often seen on genuine rough diamonds. The fake rough diamonds were intercepted by a rough diamond dealer in South Africa and extracted from the parcel. Perhaps the easiest fraud to perpetrate but the fraud with the greatest impact is when exploration samples are salted. This can be done by adding a small quantity of rough diamonds, even micro-diamonds, to a sample that will be analyzed by an independent lab. The results of the spiked analysis are used to drive investors to purchase more shares in the exploration company. Similarly, an exploration sample that is sent for analysis could be spiked with indicator minerals. Indicator minerals are so named because these minerals, although not diamond are typically indicative of diamond bearing deposits. Likewise the analysis spiked with indicator minerals can also be used to show results better than what is real and in turn drive investors to purchase more shares.

Fake rough diamond found by a South African diamond dealer in a large parcel of rough. Close up image of one of the fake diamond that shows surface features etched into the stone to make it appear more like a genuine rough diamond.

Stock Market Crime

I've been investing in the Canadian diamond industry since 1997–98, and have held stock on a few Canadian diamond companies since then. Although I have not made a lot of money on the stocks what I was able to do was to have a close look at the companies and examine their were findings, where they hold property rights, and all the other geological data they release. Diamonds are a funny commodity in that they get people thinking of vast riches, especially when it comes to diamond mines. I have had so many people ask what I think of different diamond mining and exploration companies, and junior and senior established companies, looking for any word that this company could be the next big thing. I tell them to talk to their stockbroker and at the very least check the company out themselves through the Internet. In no way do I want someone taking my word on a stock and then losing his shirt. I simply say "there are over 5,000 recorded kimberlite pipes in the world and only a handful of operating mines. You do the math". Regardless, some people just want to make a quick buck. An interesting article in Resource World magazine estimated the likelihood of a diamond exploration company coming to production stage for diamonds. A grass roots exploration company where no drill targets have yet been identified has approximately 0.5%–1.0% odds of becoming a diamond producer.[115] Still, some people seem to get caught in a buyer frenzy. They hear around the office water cooler of a good company to invest in or perhaps someone of influence touting the great riches awaiting all investors willing to get in on this company before it goes big. People take their chances if they don't follow their stockbroker's advice and check out the company. There are companies that are currently on the radar screen of securities commissions in Canada and the U.S. Both the Securities Commissions in Canada, and the U.S. Securities and Exchange Commission have suspended some diamond exploration companies from trading their stock. This is because of questionable business practices on several fronts. In more serious matters criminals have used stocks to generate criminal financing in what are called pump and dump schemes[116]. They create a company and pump up the value of the stock through aggressive marketing, public announcements and false or misleading claims. Then once the stock price is high and the scam has about run its course, the criminals dump the stock at its high values. Ultimately the scam unravels and those who held stock in this company when it fell apart are left with no value in their stock. One of the most recent and notorious companies to use false information to increase investor fervor and stock purchases was Bre-X. The misleading claims of gold findings and potential caused investors to make large purchases of this company's stock, and to push the stock value up excessively. The claims were found to be false and the stock crashed. The bottom line is with stocks one must be careful and as seen in recent years, even with a large Blue Chip stock in Canada and the U.S., there are still risks.

Proceeds of Crime and Money Laundering

There is a difference between proceeds of crime (POC) and money laundering. Money laundering was talked about earlier and has everything to do with making the criminal money or proceeds of crime appear as if it is legitimate. It's the proceeds of crime that needs to be laundered, however diamonds being a cash-like instruments, compact, very valuable, and virtually untraceable, don't necessarily need to be laundered. Diamonds, jewellery and gold are the perfect mediums for storage and movement of proceeds of crime and for generating proceeds of crime. In terms of money laundering, some academics estimate that in 2004, approximately 4.5 billion worth of laundered money may have crossed U.S. borders via the international trade in precious metals and gems[117]. Martin Frankel of the United States, who swindled over $200 million in one of the largest frauds in American history, converted millions of dollars in proceeds of crime into diamonds. When he was arrested on the run in Germany, investigators found millions of dollars of loose diamonds in his possession.[118] Others in the United States have been charged with violations under federal money laundering laws and property forfeited to the government included large amounts of diamond jewellery.[119]

The simple acquisition of diamond and like commodities is like having money in the bank. In fact according to some sources diamond has appreciated faster than inflation to the tune of 7.9% year over year.[120] But this is different and better than a bank because there are no reporting protocols that the banks require for large cash deposits. In fact, in some transactions there are no records at all and getting diamonds and diamond jewellery is easy. The cash-like qualities makes them easily traded for contraband and for the astute criminal, this can be very profitable. A criminal that takes jewellery in exchange for marihuana or cocaine makes money two fold. First, he's not selling the drugs at his cost. It is marked up in value so that he'd be making money on it anyway. If he takes in the jewellery at ten cents on the dollar to pay or exchange for the drugs, he's going to be able to turn the jewellery around for a profit as well.

In acquiring finished diamonds there is not the same urgency to make these items appear legitimate as there is for cash. This is because there are no records for police to check how much they have accumulated in jewellery and the likes, even if they keep the products at a bank in a safety deposit box. There is no monthly statement that authorities could check indicating that the criminal has a large quantity of diamonds and various other gemstones. But if something else was to give the criminal away, the police could get access to the safety deposit box and the jigs up—maybe. It all depends if the police can connect the proceeds of crime to the criminal activity.

The very nature of these items makes the transition from proceeds of crime to money laundered very easy. For instance, simply removing the stones from the jewellery and melting the gold makes the jewellery virtually unidentifiable. The diamonds could be taken or sent out of the country to some place like Dubai, where diamonds are welcomed and money laundering with diamond has been reported[121]. Movement of diamonds is really quite easy and Dubai would be a great place to sell the stones and set up a rainy day account. Even simpler, the stones could be sold to a regular buyer through Internet auctions or to large dealers in other countries providing a steady cash flow. To make the trail even more difficult to access, the buyer of the stones can open an account at a U.S. bank and then make the deposits or payments into the account. All the seller needs is for the buyer to send the seller a bank card and they're in the money, making withdrawals on that U.S. account at ATM's in Canada. The diamonds could also be sold to any number of jewellers at a discount to the wholesale price to put some quick cash in the pocket without the knowledge of law enforcement.

For criminal financing or money laundering, diamonds and jewellery overcome one of the great obstacles of criminally acquired property and wealth, which is its legitimization. Because there are no laws governing the possession or sales of diamonds and jewellery and because these items are difficult to identify and/or trace, criminals can sell them openly and move criminally acquired diamonds and jewellery back into the legitimate jewellery market undetected. The criminals are then able to seemingly legitimize their proceeds of crime in what is known as money laundering. And in many countries, when it comes to diamonds, law enforcement and Customs have limited understanding and capability to effectively investigate and control money laundering[122].

The use of jewellery stores as facilities to launder money is not new and has been used by large organized criminals such as Pablo Escobar, head of the Medellin drug cartel.[123] Governments are reacting to this and in the interest of truncating the money laundering capabilities of organized crime and uncovering terrorist financing operations they have been tracing cash transactions exceeding $10,000. The United States Treasury has labelled diamond merchants, jewellers, and related industry outlets in a similar manner as financial institutions. Therefore when transactions exceeding $10,000 are made through the sales of diamonds and diamond jewellery, businesses are required to report these transactions as if it were a financial transaction exceeding $10,000. Canada is examining similar legislation that would require those who operate in the diamond and jewellery industry to comply with reporting protocols similar to those that are already established for financial institutions.

This is a noble action but not without its potential failures. With a transaction of $10,000 dollars cash it is simple to record, track, and confirm in fact that these moneys were transferred from one party to another. This is because there is no subjectivity to the transaction, one dollar equals one dollar. In addition the movement of cash is increasingly an electronic transaction which is recorded, stored and can be easily examined. On the other hand, cash-like instruments such as diamond present a much more difficult monetary instrument to accurately record, track, and confirm values. A $10,000 purchase of diamond for money laundering purposes at a retail jewellery store will not translate into $10,000 cash in hand. The retail values, wholesale values and liquidation values are key to deciphering who is money laundering and where it is taking place. In addition, these transactions are not recorded in financial institutions. Confirmations and transaction histories are only laboriously uncovered through physical examination of inventories and documents. Also, stockpiles of diamonds can be quickly accumulated outside of an inventory ledger through smuggling and off-street purchases. There is also the historical practice of diamond trading for cash, without papers or record of transaction, which is nearly culturally ingrained into some diamond trading circles. Although this practice is slowly diminishing, together these issues represent hurdles to the creation of effective legislation respecting transactions of diamond and like commodities.

The most difficult obstacle to overcome is that of the values of diamond because assessing the value of a diamond is subjective and easily manipulated. For cash, the value of a dollar is fixed aside from market fluctuations. Diamonds' value is based on four separate valuators, market fluctuations based on supply and demand, and international currency rates. The value is assessed by experienced diamond graders who make estimations of the value of the stone based on industry information and guidelines used to grade the quality of the stone. While everyone knows the value of one dollar in cash, there are likely few people in any government agencies who could place a value on a diamond. This is important knowledge to have if one is creating tactical or strategic intelligence on this crime typology. As indicated, the valuators used to establish a diamond value are in fact guidelines, and are not fixed and vary from person to person and country to country, making it impossible to establish an absolute price for a diamond. The very nature and subjectivity of diamond valuations is the centre point of criminal exploitation in nearly all aspects of the criminal use of diamond. As such it is quite conceivable that through manipulations of value, a person could record and report the sale of $10,000 worth of diamond that could actually be worth up to $30,000 or the reverse could hold true and a sale of $9000 worth of diamond could go unreported yet have a true value of $27,000. This means that a description of the item is as important as knowing what the item was sold for. Ultimately, a financial report that doesn't include a description of what is being sold would be similar to a financial transaction report of money that doesn't state what currency it was made in.

Inventories and values are easily manipulated as is evident in cases of tax evasion through undervaluation, money laundering through overvaluation and conversion that occurs with criminal use of and importation of diamonds and precious gemstones. I have seen diamond dealers that deal several million dollars worth of diamonds per year catalogue their inventory on a scratch pad. Inventory records such as these would impede actions by authorities investigating money laundering crimes. Cases involving diamond value evidence could be difficult to prove as the subjectivity involved in establishing diamond values could be used to refute crown evidence in court. In addition, inventories are often recorded by carat weight and not the number of pieces in the parcel, such that a business may show it has 100 carats of diamond in inventory. This 100 carat weight could be composed of 100 diamonds as easily as it could be composed of 1,000 diamonds of the same quality, reflecting different individual stone weights and having totally different stone values. The recording of diamonds based on a parcels carat weight and not with a corresponding number of pieces in the parcel represents a loophole experienced in current legislations such as the Customs Act and the Export and Import of Rough Diamond Act. False receipts, cash buying of diamonds, undervalued imports of diamond, and smuggling can be used to skirt any reporting protocols that may be established. In addition, black market purchases and buying diamonds off the street also assist with inventory manipulation, this in turn allows for valuation manipulation and the ability for criminals to overcome government regulations. With criminals dealing in diamonds there is black market for diamonds in every major city in Canada and one report suggests that over half of the diamonds over 0.50 carat in one Canadian city are sold through the underground market.[124] This ability for criminals to acquire diamonds off the radar also allows them to defeat government controlling measures.

Effective money laundering legislation targeting gemstones and precious metals requires measures akin to checks and balances which ensures reporting captures the salient information in any transaction and that valuations are not being manipulated to undermine or skirt the reporting protocols. With Canada having taken over the presidency of the Financial Action Task Force (FATF) in 2006, there is an even greater need to show that Canada has taken a lead role in curbing criminal and terrorist use of diamond. This will result in the creation of anti-money laundering legislation focussed on the diamond, gemstones, and precious metals dealers. To be truly effective, the use of accounting procedures and examination of inventories by qualified personnel may be required. However, the reality is that diamonds and like commodities and they way they are handled and dealt is completely different than that of the financial sector. As such, the template of existing legislation for the financial sector cannot simply be superimposed over the diamond, gemstone and precious metals sector. The jewellery industry is broad and among others includes retailers, manufacturers, importers, wholesalers, pawn shops, auctions, the internet outlets, and refineries. The retailers alone may number over 3000 however there are no statistics on this and these are industry estimates. The inclusion of the others could easily double the number of businesses that fall under this umbrella. This is a mammoth undertaking and gaining compliance of all those involved will be difficult. However, criminal use of diamonds and like commodities is a reality and measures are required to combat this. Those in the jewellery industry will have to comply as this is the post 911 world, and there will be no legal option. On the other side of the coin, legislation must be comprehensive, encompass all corners of the jewellery industry, and terminate all possible criminal opportunities. William Fox, head of FINSEN in the U.S., in addressing the diamond industry in Dubai said criminals seek out and exploit the weakness in any regulatory system.[125] This underlines the importance of comprehensive diamond, gemstone, and precious metals anti-money laundering legislation.

Overvaluation of Diamonds to Launder Money

For years criminals have used soaps and fertilizers for overvaluation schemes,[126] but the exploitation of the diamond's value and its cash-like qualities may make these schemes and the disposal of diamond quite easy. Not only can the diamond be utilized to facilitate money laundering of proceeds of crime, but diamonds can be sold at a profit or used in other schemes. Converting money derived from criminal activity to diamonds is a simple money laundering process as the conversion process already provides excellent insulation from authorities and typical money tracking procedures. In converting the diamonds back to cash there is likely to be a loss of value but this is very much subject to the individual, their situation and where they purchased the diamond on the mine to market chain. This will be examined at greater length in Chapter 8 "Diamonds as Currency". Even if he purchases the diamonds at full cash wholesale prices, then sells them at a 25% discount to this, these losses are within acceptable losses of the money laundering process[127]. In addition, he may be able to purchase the stones at a wholesale price or discount to wholesale and trade the diamonds for contraband. Depending on whom he trades with and what the commodity is, not only will he profit from the sale of the contraband but he will likely also have turned a profit on the sale of the diamonds. There are several ways that this can be done. For instance, if a criminal is involved in selling diamonds and trafficking cocaine he would want to be able to legitimize all that drug money. For the diamond end of the business he could buy 500ct, I color, SI clarity quality diamonds, each in the 0.50 carat range. Now based on the wholesale values of 2005 these stones would be priced at about $2000/ct for a total of $1,000,000. Because of the large volume of stones that he is purchasing, he receives a 30% discount to the rapaport price and the stones actually only cost him $700,000. There are several countries around the world where this type of scheme could take place. Some countries are gaining a reputation for the money laundering that is done through its diamond trade. Reports indicate that although diamonds are simply flowing through a particular country with no added work, sometimes their values have doubled upon export.[128]

> 500ct of diamond x $2,000/ct = $1,000,000 less 30% discount = $700,000.

To fool everyone involved and to facilitate the money laundering process the criminal obtains a receipt for $2,000,000 stating the diamonds are of H color VVS clarity quality diamonds that would sell for approximately $4000 per carat based on full wholesale value and with no discount to the price.

> 500ct of diamond x $4,000/ct = $2,000,000.

Therefore the diamonds entering Canada come with a receipt and a stated value that appears accurate and taxes paid to the tune of about $120,000.

> Diamond cost $2,000,000. x GST 6% = $120,000.

The total amount that has actually been paid for the stones is $700,000 plus the taxes of $120,000 for a total of $820,000. This is the total cost of the diamonds.

Even if the diamonds are held and sent for analysis/grading, the worst thing that could happen is that the diamonds may be found to be worth less than what is stated and the individual ends up paying less tax. As discussed, putting a value on the diamonds can be a very ambiguous process, as several of the stones have to be graded to give a representation of what the parcel value likely is. In a 500 carat parcel composed of mostly 0.50 carat stones there would be about 1,000 diamonds in the bag. Roughly 10% of the stones would have to be graded to establish a ball park value for the parcel. Even more would have to be graded to be precise. In a parcel with a variety of diamond sizes a larger number of stones would have to be sampled and the stones would have to be separated. This is certainly not something that's going to be done on a whim. What usually unfolds is if you have a receipt for the stones and they look legitimate then the authorities simply assess the taxes, stamp the papers, and move on to the next person. Customs officials do not have the skills to make a preliminary analysis or the time that would then lead to a full gemmological analysis. Again, the criminal exploits law enforcements lack of knowledge of this commodity.

The diamonds are now imported into Canada and the criminal has paper stamped by Canada Customs that legitimizes the values that he has put on the stones. (Likewise, when rough diamonds are laundered through mixing or switching gems or parcels and Kimberley Process certificates are issued for the parcel, the KP certificates provide added legitimacy to the diamonds[129].) The company now enters the finished diamonds into inventory and shows the $2,120,000 costs for the stones on the books. The reality is the stones only cost $820,000 and this allows $1,300,000 for money laundering purposes or in this companies case, to launder the drug money profits. The company can now over-grade the stones and sell them off at retail venues in Canada or internationally at a further financial profit.

Tax Evasion

Until recently, undervaluation of diamonds was especially criminally lucrative in Canada because a low value for duty meant that the importer paid less excise tax. In 2003 the excise tax was 10% on diamond imports. It was decreased to 8% in 2004, to 6% in 2005 and then eliminated in 2006. Lowering the import value of the diamonds meant less excise tax and less GST (Goods and Services tax of 6%). With the elimination of the excise tax, one of the benefits to undervaluing diamonds is found if the criminal sells the diamonds without GST or in reducing the amount of tax payable in the case of a personal (non-commercial) entry for diamonds. The reason for this is because when a company imports diamonds they have to pay the GST (tax) on the diamonds. When they sell the diamonds they also have to collect the GST again from the consumer. However, when they pay the government the GST they collect, they also get back any GST they have already paid to acquire the diamond and this includes the GST paid on the import of the diamond. For example, an importer brings in a diamond from the U.S. at a value for duty of $1,000 Canadian. He will have to pay 6% GST or $60 on the diamond. He then sells the diamond for $2,000 and collects $120 GST, which he has to pay to the government. However, he gets to deduct the GST he had paid on the diamond on import so he actually only has to pay the government $60. As a result, for tax evasion purposes, if the company is remitting the GST they collect to the government then there is little incentive to undervalue the stones to reduce the import GST because ultimately they are going to get the import GST back anyway. On the other hand, if diamonds are brought back to Canada on a non-commercial entry, for personal use, there is still the incentive to undervalue the diamonds to lower the GST because in this case they will not be able to recover this tax from the government. In other countries around the world, there are varying amounts of import taxes and different taxation systems. Depending on the amount of tax and type of taxation system, under-valuation of diamonds is still very lucrative. Cambodia, for instance, has a 35% import tax on diamonds. Using Canada as an example, the benefit of undervaluating diamonds is easily seen. While it was even more lucrative in the days of Excise tax on diamonds, it is still lucrative today to undervalue diamonds, especially for the criminal that sells the diamonds on to the consumer without collecting the GST on the final sale.

When making purchases overseas or south of the border, international exchange rates could make it even better for criminals or those who walk the line to want to smuggle. This is especially so when the Canadian dollar drops relative to foreign currencies and for two reasons. First, the exchange rate means paying higher relative taxes on the product to bring it into Canada and secondly margins are squeezed even tighter because of the poor exchange, which can make the idea of smuggling even more appealing. In these times of dollar fluctuations, the increased costs create a margin squeeze that can be eliminated through raising the retail price of the products or through smuggling and/or undervaluating the stones to evade taxes.

Large stores, especially chain stores, have bulk buying ability and greater inventory of stones that can somewhat insulate them from price fluctuations. This creates a situation that could lead to smuggling stones or undervaluation to evade the taxes for those who don't have these same competitive advantages. In smuggling stones, the criminal still has to pay the higher exchange on the dollar to purchase the stone but through smuggling he skips the taxes and the ultimate cost of the stone stays similar to when the dollar was trading at a higher level to the US dollar. I remember talking to Canadian wholesalers and retailers a few years back when the dollar was trading at the 61–62 cents level to the U.S. dollar. Many of them said they were losing money while others were barely profitable. Having said that, and noting the Canadian dollar is presently trading in the range of 85 cents to the U.S. dollar, the situation must be better. That is because the same $3000 dollar U.S. priced diamond they were buying a year ago at an exchange rate of $1.43 is now being purchased at $1.16 Canadian to the U.S. dollar or $3480 Canadian. Even with the GST taxes of about $208, the stone cost less than it did when the exchange rate was $1.43 Canadian. This must be clarified somewhat because the conditions stated here change over time and there are more things at work than simply the Canadian dollar exchange on the U.S. dollar. The U.S. dollar exchange against the Euro is also very important because if it drops against the Euro then the price of U.S. imported diamonds goes up, which in turn raises the prices of the stones that Canadians import from the US. Usually these particular monetary conditions that come together to create an environment conducive to smuggling are short term windows of opportunity that span only a few months until inventories turn over and prices adjust.

There is also the prospect of criminals failing to declare work done on diamonds. For instance, a buyer of fractured, chipped, or otherwise damaged diamonds may send the stones out of the country to be re-cut. There are several companies that offer this service at recutting fees in the neighborhood of $100/ct. For instance, a criminal sends a 1ct stone to the cutters with a sizable chip out of the girdle. He bought the stone for $600, a fraction of its price if it were not damaged, then sent it to the U.S. to be re-cut to remove the fracture and the result is a 0.75ct stone worth 3 times the value of what he paid for it. He only paid $100 to have the stone fixed but on the return to Canada, its true value for duty is $1,800 yet it clears Customs as $700 and he evaded the tax on $1,100. In any case where work is done on a diamond that enhances its value, the increase in value is then subject to taxes. Criminals who know this will not report the true increase in value of the diamond or the work that was done and in doing so can exploit the system.

Undervaluation to Move Money

It is with increasing difficulty that criminals are able to move money across borders. Criminals can exploit the subjectivity of diamonds value to move money across borders far more effectively than cash. Quite simply a parcel of diamonds that is worth $500,000 dollars or any amount for that matter, can be sent to another country as being worth half that or less. The diamonds are like cash and while the authorities may recognize the amount that is being declared in the cross border transaction, the reality is the recipient has received far more "cash" than the authorities realize. This type of arrangement can be disguised as a business transaction and without further information would not draw anyone's attention. This method of moving cash can be used to pay for contraband and to move proceeds of crime.

Smuggling

Similarly, smuggling of diamonds can also be used primarily to move money across borders or to evade taxes. In jurisdictions where taxes on imports (or exports) are high you could expect smuggling to occur as a way to eliminate or evade the tax. Sergeant John Philpott of the RCMP sums it up best by saying, "If an analyst were to plot the specific import/export taxes of any particular commodity, country by country, the potential smuggling routes would emerge." At least insofar as tax evasion is concerned.

While diamonds are likely the easiest cash like instrument to smuggle, this method presents increased jeopardy to the criminals, relative to those who would simply undervalue the diamonds, in that if they happen to get caught they will loose the stones and face criminal charges. Smuggling diamonds is simple. They are easily hidden, typically do not show up on x-rays, by metal detector, or by police sniffing dogs. Conceivably, $1 million dollars worth of diamonds can be sewn into clothing or inserted into electronics and then transported around the globe.

But smuggling diamonds is better than smuggling cash or not reporting the international movement of money. If you try to bring over $10,000 cash into Canada and you do not declare it, it will be seized by Canada Customs (C.B.S.A) if they find it. This is a function of the money laundering anti-terrorism legislation. In fact between January 2003 and August 2007 Canada Customs seized approximately $145 million dollars in over 5800 separate events[130]. With diamonds there are at least a few legal options that the criminal can try to employ to get the stones back, but if the criminal smuggles cash, it is all but gone.

Misdescription

This is not unlike smuggling in that it is a magnificent way to move money internationally but it involves actually declaring the product to Customs of whatever country the diamonds are being sent to. Simply, declaring that the parcel is white sapphire, topaz or one of the diamond simulants with the appropriate lower values of the gemstone can enable a criminal to move a large amount of money via diamonds over borders while reporting a tiny value to the authorities. For instance I could commercially import 10 x 1 carat diamonds into Canada with a true wholesale value of $50,000 but label the parcel as 10 x 1 carat white sapphires at approximately $1000 wholesale. A great way to move money, but in addition this is another way to skirt the taxes. In this case approximately $3,000 worth of taxes would be evaded.

GST Fraud

The GST (Goods and Services Tax) is not popular. Regardless, the reason the tax is collected is for government coffers to provide revenue for national programs and service these departments. Having said that, these taxes are collected on products and services imported into Canada, sold in Canada, or provided in Canada. The government does not collect these taxes on products that are bound for other countries and for this reason any GST that has been paid on products in Canada can be reclaimed from the government when the product is exported from Canada. GST fraud is not new to the government. People have been doing this since the inception of the GST. The fraud is simple. A product is imported into Canada and GST is paid on the value for duty. The importer retains the receipts and sells the products under the table in Canada for profit and without charging GST. Some time later the importer sends a parcel or container out of the country with inferior or worthless items that have been described as those same goods that were earlier imported. Like most products leaving the country, no one examines the shipment. When the products enter the destination country, if they are not contraband and are similar to what the export documents say, then they are admitted into the destination country. The fraud artist obtains export documentation from Canada or receives paperwork from the destination country that shows the products entered the destination country. He then sends this paperwork to the government with the paperwork that shows the GST he originally paid. In a few weeks he gets a refund for the GST he paid on the products he originally imported—very simple. The potential for this type of fraud and the criminal benefit is even greater for diamonds than nearly any other product because hardly anyone knows diamond, it value, or for that matter can even identify diamond. So the person who imports a diamond, pays the GST, then turns around and sells the diamonds under the table and could go ahead and export what he declares to be the same diamonds. Maybe what the criminal exports is a CZ (Cubic Zirconium) worth about $5. He obtains the Canadian export documents or entry documents (also known as Consumption Reports) from the destination country and sends this documentation to the Canadian Government with the original receipt to claim back the GST. To give an example, a $5,300 diamond imported into Canada with full taxes paid returns $300 GST taxes back to the fraud artist. In addition the original stone that was imported is kept off the books and is then sold under the table without GST or HST (Harmonized Sales Tax). Further, there would be no income tax to be claimed on profits from the sale of the stone. In this case of fraud, the GST reclamation fraud perpetrates other frauds in order to make the transactions look legitimate and to give the appearance that the inventory and ledger books are straight.

Courier and Mail Movement of Diamonds

Getting diamonds in and out of Canada is easy. Like any other product and with the help of open markets and free trade, an export document is completed declaring what is in the parcel, with similar processes for importing products. I have imported hundred of diamonds and to the credit of the Canadian Border Services Agency, several of the parcels have been opened and examined. Diamonds are regularly moved through courier services around the globe. A few chapters back I mentioned rough diamonds that the two dealers in the U.S. were prepared to send to Canada via courier and in the last paragraph that diamonds are easy to move. The reality is that courier services are being exploited every day by criminals to move diamonds and more. Steroids, cigarettes, drugs, diamonds, and gold are regularly shipped unwittingly by couriers over borders and within Canada to escape taxes or to be moved undetected. Any large courier service has thousands of parcels coming into Canada every day and they all must be cleared by Customs. It is neither practical nor possible to examine all the parcels and CBSA target parcels randomly or ones that are unusual. They are further break shipments down into those that are over or under $1,600 dollars. Those over $1600 are what is known as high value shipments and are subject to higher scrutiny. The fact that they have a $1,600 value also means that they likely have something being imported that requires taxation. Most letter posts sent into Canada enter in a small cardboard pack that is very thin are usually carrying papers and are declared at values under $1600 dollars, a low value shipments. Someone could courier one of these letter post cardboard parcels, with an ounce (approx 150 carats) of diamonds, into Canada labeled as documents or birthday card and there is a chance it could go right on through. Remembering the email I got from the rough diamond dealer in the U.S. This is exactly how they were going to ship the parcel, misdescribed and without a value attached, to make it appear as a low value shipment.

Theft from couriers is not uncommon. If a criminal happens to recognize an address or company name as being in the jewellery industry, if the opportunity arises he might take the parcel thinking there are big riches inside. Most people in the business know this and as such disguise their company names but addresses are necessary. Sometimes disguising the company name is not enough. Theft of parcels in transit has happened even to Canadian diamonds. In 2000, Sirius Diamond Cutting facility had a parcel of cut diamonds stolen in transit from Yellowknife to Vancouver.[131] With some couriers if a shipper chooses to purchase insurance on his parcel, the courier company will put a great big sticker on the parcel that discloses to every one how much the combined shipping and insurance costs are for the parcel. If the parcel is very small, and light, and has a sticker on it showing that the cost to ship the parcel is very expensive, it is pretty easy to make the determination what is inside. Anyone can do the math and figure that the most it would cost to send a small parcel is about $10 dollars anywhere in Canada, even through expedited service. But if the sticker shows the cost to be $30 or $40 dollars, it is obvious that the bulk of the cost is the insurance. It is that easy, if the parcel is small and has a big sticker price of shipping and insurance costs and when you shake the parcel you can tell something small is rattling around inside, the rest can be figured out. This is certainly a soft spot in the security and at least one courier has experienced several losses of parcels as is evident in the fax fan out I received from a supplier (see below). The industry alert suggests buying insurance for the parcel, however based on the example above, it may be better for companies to get insurance through a secondary party like the Canadian Jewellers Association or Jewelers Mutual so a high sticker price is not reflected on the outside of the parcel.

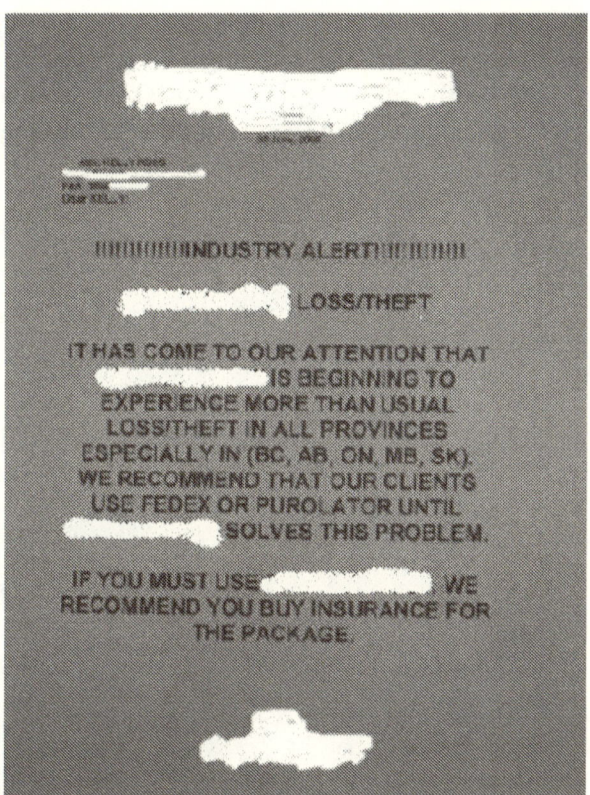

Diamond Trade for Contraband

There is evidence that diamonds are increasingly being traded directly for contraband products in a convenient criminal effort to short circuit government efforts track money movements. One study out of Australia showed that 70% of criminals had traded stolen property for drugs[132]. Additional research suggests this is a relatively recent phenomenon of the 1990's[133]. Newspapers and government reports, and several books suggest that diamonds, gemstones, and gold are being regularly traded for contraband. This alternate currency potential has already been examined in depth with respect to blood diamonds in West Africa being traded for weapons, ammunition, and explosives. Newspaper articles and document searches show the trade of jewellery for drugs is prevalent in the United States, Australia, and New Zealand.[134] This makes perfect sense for the world of criminals. Diamonds and jewellery are cash equivalent items and can be concealed and carried across borders without detection. There is no need to store the diamonds in a bank and they are easily hidden from other criminals. They provide a convenient alternate means of transacting a weapons or drug deal. Criminals are able to do this because the drug dealer knows that diamonds are valuable and he will be able to get a large quantity of cash when he sells the diamond.

7

Treatments, Scams and Misrepresentations

Treatments

Laser Drilling

For as long as man has cherished diamonds, the diamonds that were the top colour and clear of inclusions have been highly prized. Books and transcripts from India dated from 300 BC to 500 AD describe a clear diamond that has a well-formed crystal shape as a quality that makes diamond valuable[135]. As can be imagined, black inclusions in a diamond stand out like a sore thumb and detract from the beauty of the stone. In modern times, clever diamond sellers crafted ways to reduce the appearance of inclusions in a diamond. One of the most common methods is a treatment process known as laser drilling. This process was developed in the 1970's as the application of laser technology was gaining momentum. This process involves using a laser to literally drill a hole from the surface of the diamond to the inclusion inside. Once the hole is drilled, strong acid is used to dissolve the inclusion or bleach it to a white color, which then makes the inclusion less visible to the naked eye. The drill hole is filled with a clear resin that hardens and the procedure is complete. The result of the treatment can raise the clarity grade of the diamond by one or two clarity grades perhaps as much as an I1 to SI1. Of course like all diamond treatments that change the appearance of the stone from its natural state, this treatment must also be disclosed. There are limitations to this treatment in that the stone can only be drilled to approximately 1.6mm. Inclusions that are deeper into the stone may not be able to be treated this way. On top of that, the drill channels are visible under 10x magnification so they can be seen through a 10x jeweller's loupe or microscope.

In recent years a newer laser drilling process has emerged that is even harder to detect. This process can reach deeper into the stone and is more difficult to detect. Some diamond industry specialist suggests that many laser-drilled diamonds are sold without full disclosure that the diamond has been enhanced with this treatment. I have only seen a few laser drilled diamonds in the thousands that I have examined. However, this still presents an opportunity for criminals to scam the general public and generate profits by selling inferior or treated diamonds at prices commensurate with higher quality, untreated diamonds.

Fracture Filling

Imagine someone buys a great looking diamond with no apparent fractures or flaws. All the info about the diamond was received before purchase and the buyer believes he made an informed purchase. Several years later when having the claws repaired on the ring, the jeweller advises that the diamond has been enhanced and there is a large fracture that cuts through the stone that has been filled. Then the jeweller explains about fracture filling, the process of filling fractures in diamonds to enhance the clarity of the stone. Criminals can sell clarity enhanced diamonds through jewellery stores, auctions, or the Internet and they will do so without disclosing the treatment.

This treatment emerged in the late 1980's and is not unlike laser drilling in that it is designed to improve the clarity of the diamond. The object is to fill in cracks or fractures that emanate from the surface with a clear resin so that these fractures become nearly invisible. Once the resin fills the fracture it is heated to 400 degrees Celsius which sets the resin and the process is complete. Like laser drilling, this process can raise the clarity of the diamond by one or two clarity grades but because the resin is not completely colourless it can also lower the colour quality of the diamond by one grade. This treatment can be detected under magnification as there may be bubbles trapped in the resin. Also, the resin may produce a pinkish or greenish iridescence when viewed on an angle under magnification. Of particular concern is that this treatment will deteriorate over time and is not permanent. This process should be disclosed to a buyer, however, an unsuspecting buyer could quite easily purchase a fracture filled diamond and never learn of it until they tried to sell the diamond. This is another opportunity for a criminal to commit a fraud.

HTHP Treatment

High Temperature High Pressure treatment of diamonds is a relatively new way of treating diamonds to increase their colour quality and provides great potential to defraud a buyer. The process involves subjecting diamonds that are literally brown in colour to the same temperatures and pressures experienced in diamond formation. This brown colour in some diamonds is believed to be caused by poor alignment of the crystal structure. This can happen to a diamond if it has been subjected to the Earth's forces that literally squeeze the normal crystal structure into something irregular. This is known as plastic deformation. Subjecting the diamond to the HTHP process under controlled conditions allows the crystal structure to realign itself and the brown colour disappears. The results show that this treatment can turn what was a brown coloured diamond to a diamond with a white, top white, or even pink colour[136]. Again like any process that enhances the natural diamond, it must be disclosed to a buyer. However, detecting HTHP treated diamonds is quite difficult for the average gemmologist unless they have the necessary equipment and know what to look for. Basic equipment is available which can lead a gemologist to a suspicion that a diamond may have been treated this way. As this process gains momentum there will be a lot of these diamonds entering the market through criminals while the fact that it has been treated will undoubtedly go undisclosed and undetected by the public.

This treatment process can be difficult to detect and only the most sophisticated gemmological labs have the equipment to positively identify a diamond that has undergone the HTHP treatment. While these treated diamonds are legitimately sold and often laser marked to denote the treatment process, the potential exists to use these in frauds.

Irradiation

I was on a Caribbean cruise a few years back with several friends and got pulled from one jewellery shop to another to, as my friends would say, "make sure we got a good deal". I already had explained to them that I'm not going to stand in the shop and debate the qualities of a stone with the vendor and if they really wanted a good deal, that I would help them out when we got back to Canada. Truth be told I don't mind spending a day checking out jewellery shops, so I was quite happy to tag along. As I have learned, for many people, it's not necessarily getting the best deal that counts but getting a good deal and a fun memorable experience. I think this is one of the drawing cards that these Caribbean jewellery shop have with the tourists. First, the tourists have the impression that they are getting a great deal, usually because most of the advertising signs say duty free or something similarly enticing. Second, in getting this supposed good deal, there is the memorable experience of latching onto this great jewellery piece while on vacation. There are few items a person can bring back from vacation that carry this same appeal to an owner that is also actually used and does not end up in a garage sale.

Imagine the let down if this 'great deal' and wonderful memories were tainted upon return to Canada and learning that what you were told you were purchasing was a lie. That's what would have happened to a friend of mine on the Caribbean Cruise. We'll call them Ken and Jane. Ken and Jane were looking for jewellery keepsakes of the holiday and Ken had found a tension set platinum ring set with what was described as a natural intense blue 0.20 carat round brilliant cut diamond. When Ken asked if I would come check it out I was quite interested and we had a look. The ring was quite nice and the stone well set and it fit Ken's finger nicely. However, there was a problem with the stone and the price. The stone was very clean and a nice rich blue with good saturation of color throughout the gem and very bright. The ring was priced at approximately $900 U.S. and although I was unaware of the true cost of such a stone I suggested to Ken that for a natural blue diamond it would be significantly higher. The ring was priced accordingly if the diamond in it were genuine SI2, H color round brilliant quality. In addition, natural blue diamonds are extremely rare and would likely not be found in cruise boat. I asked the clerk about the ring; if the diamond was a genuine untreated diamond, which he said it was. In private I suggested to Ken that the clerk probably didn't know if the stone was genuine untreated diamond. Then I asked the clerk if the diamond came with a certificate of authenticity, which such a rare diamond should. There was no certificate and it was my impression that the stone in the ring was in fact an irradiated diamond. Without lab examination it would be difficult to say for sure but in this case I told Ken to follow the old pricing rule about being too good to be true.

Irradiated blue, yellow, and green diamonds have been popping up on the market now for several years. Of course natural diamonds in fancy colours of blue, yellow, and green are extremely rare and command the highest of prices for diamond. A 1.00 carat natural fancy colour pink diamond can easily cost over $100,000 dollars. To the unsuspecting buyer, an irradiated diamond would appear very much like a genuine fancy coloured diamond but could be priced very much below what a genuine untreated stone would cost or it could be priced the same and sold as a fraud. Without disclosing the treatment the stones could then be sold as genuine.

Scams

Diamond Shorting

Have you ever been to a bar and ordered a shot of whiskey? The bartender usually pulls out a 1 ounce glass and fills it to a pre-determined line on the glass indicative of 1 ounce of liquid. Some bars have dispenser that automatically measure out 1 ounce of the liquid. Every so often the bar tender is short on the ounce and doesn't fill the glass right up to the line. Similarly, the automatic dispensers are calibrated to measure an ounce and over time can become inaccurate. Some bars do this purposely because over time the amount they save in shorting people on their alcohol adds up to big savings for them. So the question then is what determines an ounce or better still, how close does the bar tender have to be in pouring that shot of whiskey so that it is still considered one ounce of whiskey? For whiskey and nearly every other liquid, solid, or gas that is sold by weight or volume, there is a law that governs these amounts. More specifically the law, known as the Weights and Measures Act, governs the units of measure for nearly all products as well as the tolerance in these measures. The tolerance is basically the amount that a measure of something can be incorrect and still be considered the full amount being measured. For instance, that 1 once of whiskey could perhaps be off by 1 millilitre yet still be considered 1 ounce. Perhaps that 1 pound streak you purchased from the butcher could be off by 5 grams and still be considered a 1 pound steak. These numbers are arbitrary and for example only, but this explains tolerance on a product.

The same rules of tolerance hold true for diamonds. The tolerance for diamond is perhaps erroneously grouped in with "Limits of Error for Precious Metals and other Commodities of Comparable Value the Quantity of Which is Stated in Metric Units of Mass" under Schedule II, Part XV of the Weights and Measures Act Regulations. We know that gem diamond is in no way similarly valued to diamond as 1.00ct weight of low quality gem value diamond could easily sell for $2,000 U.S. per carat. On the other hand 1.00ct weight of pure gold (aka 24karat gold) would sell for about $3–4 U.S. per carat.

Under the Weights and Measures Act, the tolerance for Precious Metals and Similarly Valued Products, which includes diamond, is approximately 12 milligrams for 200 milligrams. To convert this, 200 milligrams equals 1.00 carat weight and 12 milligrams equals about 0.06 carat. So this is actually a very liberal tolerance because this means a 1.00 carat diamond could be out as much as 6 points or 0.06 carats and still be considered a 1.00 carat diamond. This is the law, and it is quite frightening because legally a criminal could sell you a 0.95 carat diamond as a 1.00 carat diamond. In the Canadian Guidelines with Respect to the Sale and Marketing of Diamonds, Coloured Gemstones and Pearls (revised edition 2003), it lists that the proposed changes to the limits of error would be a 0.002 carat limit of error (tolerance). However, these changes have not yet been made to the Weights and Measures Act as of June 2007.

This presents a hornet's nest of problems, not the least of which is that many people purchase 1.00 carat diamonds solely because the size of 1.00 carat is seen as a benchmark. Imagine the surprise, then disappointment, then anger to find out that they were shorted on their 1.00 carat diamond and in fact their diamond is not 1 carat but less, somewhere between 0.94 carat and 0.99 carat. This loophole is an incredible opportunity for criminals to legally use diamonds in a scam or legally, as a measure of weight to sell diamonds.

As mentioned in the chapter on diamond valuations, diamond values are not linear. If valuations were linear, then for example if a 0.25 carat diamond is worth $400, then a 0.50 carat diamond of the same quality would be worth $800. The reality is the 0.50 carat diamond is more likely to be worth about $1,200 showing the non-linear valuation scale of diamonds. This is because the larger the stone the higher the relative rarity of the diamond, which in turn causes the price per carat to jump from one weight class of diamonds to the next. This is seen in the Rappaport price reports and other such reports like The Guide. One of the largest price jumps between categories is found between diamonds in the 0.89ct–0.99ct weight class and those in the 1.00ct–1.49ct weight class. For an example of this price jump, simply compare the value of a 0.99ct D colour, IF clarity round brilliant diamond with the value of a 1.00ct diamond of the same quality. The 0.99ct diamond is worth approximately $10,000 U.S. and the 1.00ct diamond is worth $17,000 U.S. This massive $7000 increase in price is based on the 1.00 carat diamond weighing only 0.01 carat more than 0.99 carat stone. Remember, this is wholesale pricing. The price jumps are not as dramatic in the medium quality stones but still significant. In a 0.99ct H colour, SI2 clarity round brilliant cut diamond and a 1.00 carat diamond of the same quality, the difference is approximately $400 and again this is at the wholesale level. This quality of diamond, SI clarity, H colour, is of a quality often seen in jewellery stores.

Given the fact that criminals can sell a 0.99 carat stone legally as a 1.00 carat stone yet command a greater price for it, why wouldn't they? If the stone is already mounted in the ring, the customer is not going to know the difference and there is nothing forcing the jeweller to disclose this. The scam is not really new. In the U.S. this is appears to be a common scam based on the number of web sites that have information on this topic. However, because of stricter weight tolerances of the Federal Trade Commission this could be seen as an illegal practice in the United States. In Canada, this scam could allow criminals to short diamonds to customers at huge profits. The criminal can buy a 0.97 carat diamond at several hundred dollars less than a 1.00 carat diamond and sell it as a 1 carat diamond and either make large profits or severely undercut the honest jeweller, while all along scamming the public.

This loophole could carry incredible ramifications for import and export. The liberal tolerances could be exploited to provide the icing on the cake for an import/export scam that involved switching of diamonds. The weight of the parcel going out of the country would not necessarily have to match up with the weight of the parcel coming back into the country because of liberal tolerances. The parcels would only have to be close within tolerance levels. This works to the advantage of those who are involved in cross border diamond switching scams. In fact, with diamond being so different in valuation than gold it is quite possible the tolerances used for this section could not really even be applied to diamond. It and other gemstones require a tolerance of their own.

The weight of a diamond can be estimated through measuring its dimensions and plugging the numbers into a formula. The method of weight estimation is known as the formula method, but its results can be inaccurate to several points on a 1 carat diamond. However, the tolerance on the weight of diamond is so liberal that it is reasonable to use the formula method to establish the carat weight. This method is subject to how good the measuring instruments are and further calculations that can be simply based on visual estimation. The results are somewhat less than accurate. However, if the actual weight of the diamond came within 6 points of a 1.00 carat diamond, the weight of the diamond could be said to have been estimated and bumped up and sold as a 1.00 carat diamond.

To be perfectly clear on this, not all diamonds have their weight estimated and many are weighed on an accurate scale to 3 decimal places or 1/1000 of a carat or 0.001 of a carat. I know this is the case with most Certified Canadian diamonds as well as many non-Canadian diamonds. However, when buying a diamonds, just take a look and see if the diamond grading certificate indicates the weight is estimated. When the carat weight has been estimated often the certificate will state; (Est.), or (Estimated by Formula) right on the certificate next to the carat weight. The next thing, without tipping off the clerk, is to ask the clerk if the weight of the diamond is just an estimate or if it was actually put on a carat scale and weighed. The stores I checked told me the diamond was actually weighed despite that the certificate indicated the diamond weight had been estimated by formula. In reality, the public and many of the clerks working in jewellery retail don't even know that this is how a diamond's weight is sometimes calculated. If the clerk doesn't know, then how would they ever pass the information on to the customer? At this point, it is not deception, they just didn't know. Also, just because the certificate doesn't say the diamond weight is estimated doesn't mean that the diamond weight isn't estimated. There is no requirement to state how the weight of the diamond was established.

One final word, Jewellers Vigilance Canada has been working on this problem for several years and is advocating to have the government change the tolerance of diamond to 0.002 carat. This is on par with the tolerance for diamond in the U.S. In fact, within the Guidelines with Respect to the Sale and Marketing of Diamonds, Coloured Gemstones and Pearls, this is mentioned as a proposed change to the Weights and Measures Act. Again this is another example of the industry trying to fill loopholes to benefit of the consumer and to reduce potential criminal activity.

Appraisals

There are many different kinds of appraisals and with that several different schools of thought as to what appraisal information should indicate. The best guideline for Canadian jewellers and gemmologists is found in the Jewellery Appraisal Guidelines. But these Appraisal Guidelines are not binding. In Canada jewellery and gemstone appraisers are not regulated, licensed or in requirement of specific training. For the industry sake and protection of consumers, Jewellers Vigilance Canada and the Canadian Jewellers Institute prepared this handy document with committee members from all corners of the jewellery industry including Industry Canada. The revised edition is available at www.jewellersvigilance.ca. It outlines some basic standards that an appraisal should contain respecting information, instruments to use, and credentials of the appraiser among other important particulars such as the use of Rapaport or The Guide for establishing costs. In addition, it specifies that although Insurance Appraisal is not the only appraisals that is prepared, it is the most common. The Appraisal Guidelines defines Insurance Appraisal as "A jewellery appraisal for insurance is a comprehensive description supporting an estimate of the value to be used as the basis upon which insurance premiums will be set and should be the basis of establishing the value and limit of claim settlement at the time of an insured loss". This comprehensive document provides three fundamental points to consider when making an appraisal. One is the <u>costing point</u>, which is the retailer's wholesale cost. The second is the <u>type of goods</u>, which supports higher values for highly sought after goods. The third point is the <u>total cost</u>, which takes into account the cost of doing business. However, the most important point, which few people wants to define, is what exactly the mark up should be in order to establish an appraisal value. The fact of the matter is there is no hard and fast rule for this and it is very much an arbitrary decision of the person making the appraisal. Even though, two appraisers establish that they are examining the exact same piece of jewellery with the exact same qualities, it is more than likely that they will come up with different appraised values for the piece.

Because this is truly an arbitrary process and that there is no licensing or regulatory requirements in place for making appraisals allows for criminals to exploit the belief that the appraisal value is linked somehow to the retail value of the piece of jewellery. In fact, the retail value of the jewellery piece and the appraisal value are inevitably never the same.

Look at a local store that displays the appraisal beside the jewellery piece for sale and see if the numbers on the appraisal are the same as the retail price. Believe me they won't be. Generally speaking jewellery is marked up about 100% on the actual cost and the appraisal value is usually about 150%–200% above the actual retailer's cost. Don't be alarmed at these numbers, this is really no different than other retail mark ups on clothing and consumer goods and in many cases is much lower than other industries. However, consider that you pay an insurance premium on the appraisal value and not the retail value. Imaging that you purchased a new car for $30,000 and went to insure it but had to pay for insurance premiums on a $60,000 vehicle (100% mark up). The same principal could be held up for a house or any other property typically insured. This is problematic for the insurance industry and the jewellery industry and should raise eyebrows with the general public.

So it could happen that a criminal may be selling a diamond ring appraised at $10,000 for a retail value of $6,000. A buyer comes along, sees this appraisal and is seduced by the possibility of a great deal despite the fact that the appraiser may not even be qualified to actually make the appraisal. How would the buyer know, the practice of gemmology and making appraisals is an unregulated profession in Canada. Look at any gemmologist credentials and ask, what those letters beside the persons name mean? Here is a list of some common trusted credentials seen in North America:

F.C.Gm.A—Gemmologist, Fellowship of the Canadian Gemmological Association

G.G.—Graduate Gemologist, Gemological Institute of America

F.G.A—Gemmologist, Fellowship of the Gemmological Association of Great Britian

A.G.—Accredited Gemmlogist, Canadian Institute of Gemmology

It is not paramount but it is further desireable that the person making the appraisal also carries credentials as an Accredited Appraiser. You can find out information about this through visiting the Canadian Jewellers Association web site found in Appendix B at the back of the book.

I have seen several appraisals that were so highly inflated as to be ridiculous. One piece of jewellery I observed, that had been purchased through an auction, should have estimated at an appraisal value of $3000 dollars but came with an appraisal of $9000. In addition to scamming people with appraisal values, the appraisal certificates are often so poorly prepared and contain so little information that the likelihood of using them for identification purposes is nil.

Using an appraisal value to sell a piece of jewellery is not an accepted practice according to the Guidelines with Respect to the Sales and Marketing of Diamonds, Coloured Gemstones, and Pearls. The guide states on several occasions "It is contrary to the purpose of these guidelines to use an appraisal value s a selling tool"[137]. This is the guideline, which is the used in conjunction with the Competition Act. However, as previously discussed the Guideline is not law and appraisals are commonly used as sales tools in the industry. This falls under the domain of the Competition Bureau of Industry Canada. They recognize the problem and have warned the public.[138] However these criminal actions are difficult to enforce.

Some criminals have used appraisal values to falsely report higher financial holding or collateral.[139] Other ways criminal use inflated appraisal values is in pyramid schemes. While this happens with diamonds it is more common to see the pyramid scheme using coloured gemstone appraisals. The appraisals for coloured gemstones can easily be 10x's the true value of the stone and provide a greater exploit opportunity. Using appraisals for sales purposes have also been employed by criminals pulling telephone gemstone scams and other scams.[140] In these cases, criminals target elderly people and try to sell them a worthless or low value gemstone that has an inflated appraisal. The buyers are told that the stone comes with a certified appraisal and the stones could be sold for much more than they are worth. The reality is they pay the money for the stone, which arrives shortly thereafter but is not worth nearly as much as they were told it was. Herein is the potential for criminal exploitation of appraisal values and precisely the reason the guidelines prohibits using the appraisal for the purpose of making a sale.

Nigerian Scam

These scams are perpetually popping up throughout the general public but they also target the jewellery industry. The idea behind these scams is that someone gets an email or letter from someone in Nigeria or some foreign country. Usually he explains that he is in exile from one of the war torn African nations and escaped with his large holdings of gold or diamonds, often in the tens of millions of dollars. He may purport to be a relative of some figure with government connections and even provide you a name under the assumption that the individual will run a search on the Internet and discover information confirming who this supposed relative is. Typically the problem is that he needs someone who he can trust to transfer these holdings out of the country because he cannot move them personally. He decides to send this request for assistance to someone because the individual just seemed trustworthy. If the individual decides to help this foreigner, he has then bought into the scam. The criminal then will request that you send him a sum of money usually several thousands of dollars to arrange for and complete the transaction whereby the individual will be rewarded with a large fraction of the supposed riches. It is rather unbelievable in today's world that people would fall victim to these scams but people do, and that's why they continue. In addition, while there are dedicated police resources, the criminal activity is too widespread to cracks down on all the scams. I have been the recipient of a few of these Nigerian scam requests through email. It is quite simple for them to find my email address, they look up Canadian diamond dealers on the Internet and they will get a list of them. From there every diamond dealer that has an Internet web site can be sent an email from these scam artists similar to what is outlined. In addition to email contacts by these scam artists they will at times also send out registered letters. This gives an extra air of legitimacy to the scam because a person must sign for the registered letter and those receiving the letter realize there is a definite cost to send a registered letter. A person would ask themselves who would spend the $1.40 to send this registered letter unless it was legitimate? The fact is that criminals will spend the money because it's the cost of doing business. I received one of these registered mail letters, and as the postmaster said, "You don't see these too often coming in registered mail."

The long and short of it is that scammers have been doing this for decades. Some headway has been made by law enforcement through programs like Phone Busters (www.phonebusters.com), that tackle these types of crime, but the scammers are still sending letters and emails, so they must still be making money.

In a similar scam, criminals will go to great lengths to try to scam legitimate jewellers out of their jewellery. In the fall of 2005 I had several contacts from a man who said he lived in Ottawa and wanted to buy a very high-end diamond from me. He was looking for a 1.00 carat round brilliant diamond in top white colour and VS clarity or better. I advised him of several that I could supply him with and he indicated that he wanted to pay by money order, which is fine. However, he further wanted me to quote him what the full price was in U.S. funds so that he could draw the money order from his bank account in Florida. My spider sense was tingling on this one and I remember telling my associate that if and when this money order shows up I'd have to ensure the legitimacy of it. About two weeks later I received a call from a police officer of Ottawa Metro police department. He advised that they had recovered a shipment of several phony money orders that had been crafted by this criminal. One of these money orders was made out to my business and ultimately was intended to defraud me of the diamond. Nice work by Ottawa Metro!

In another scam in March 2006, I had an order for $250 dollars worth of jewellery arrive at my business via FedEx. The order came type written on a piece of paper with a money order from Canada Post for $800. There was no reasonable explaination for this criminal to send a $800 money order for $250 worth of diamond earrings. I checked out the money order with Canada Post and sure enough it was phony. It looked like a laser copy of a genuine Canada Post money order. This criminal expected that I would cash the phony money order and send him the jewellery and the remaining $550 before anyone found out the money order was a fake. Again, the criminal sends the fake money order by FedEx so that it seems legitimate.

The image below that shows the phony money order he sent (names and addresses erased).

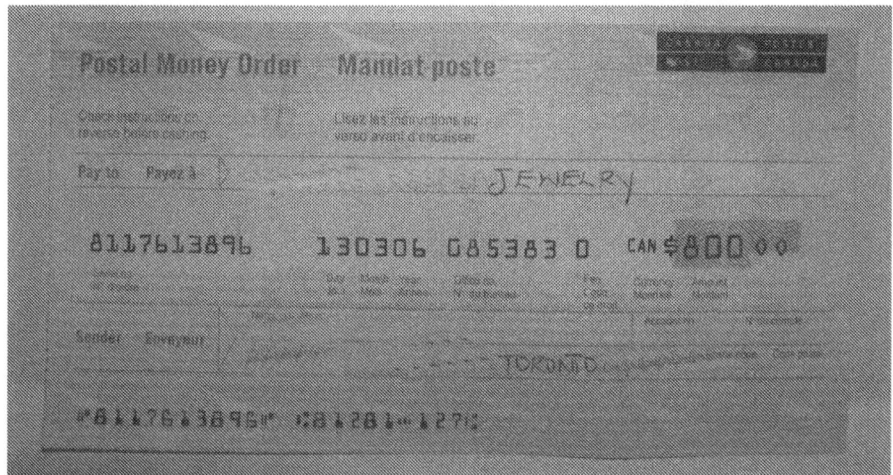

I thought about sending this criminal a cubic zirconium instead of diamonds, as fair exchange for his fake money order, but I wasn't about to waste my time, the $1.00 for the cubic zirconium or the 51 cents for a stamp, on him. Nor did I like the idea of putting a CZ in the hands of a fraud artist. This leads us to the next section.

Misrepresentations

Simulants and Synthetics

There is a myth that if the gemstone cuts glass, it is a diamond and that is how a person can tell a genuine diamond from a fake. The reality is that nearly all the gemstones used as fake diamonds will also cut glass.

Diamond simulants (a.k.a fakes) have been around probably as long as diamonds have been traded as a valuable commodity. Several natural gems that have been used to simulate diamond include white topaz, white sapphire, white zircon, glass and rock crystal (quartz). Better and often less expensive simulants were created when man became sophisticated enough to grow his own gem crystals. As a result, clear man-made spinel otherwise known as synthetic spinel became popular as did crystal and glass. These diamond simulants, especially glass, are often referred to in the industry as paste. Other synthetic gemstones include YAG (Yitrium Aluminum Garnet), Strononium Titanite, GGG (Gadolinium Gallium Garnet), CZ (Cubic Zirconium), and Moissanite which has emerged in more recent times. The most popular of these diamond simulants are the man-made gemstones Cubic Zirconium and Moissanite. Cubic Zirconium otherwise known in the industry as CZ, has been around for several decades. Other than having zirconium in its molecular structure it is not chemically similar to natural zircon. There is no natural occurrence for this gem material, it is only produced by man. This gem material has some qualities to it that is similar to diamond including its clear crystal, high dispersion and brilliance, and finally its single refraction.

Refraction is not something that will be discussed in great detail in this book. It is better to read about it in Peter Read's 1984 book "Gemmology". However, to understand the value of this quality in a diamond simulant, I'll explain it.

When gemstones are created in nature they form into two main crystal types, singly refractive and double refractive. To be a crystal that is singly refractive means that a beam of light that enters the crystal remains a single beam of light. The only crystal system that is singly refractive is the cubic crystal system. Gemstones that form as a cubic crystal structure will be singly refractive. The cubic shape is the way that diamond forms and this means that diamonds are therefore singly refractive. Other gemstones that form in the cubic crystal system and are singly refractive are garnet, spinel, and a few other gemstones.

All other crystal systems are doubly refractive and this means that when a beam of light enters these gemstones, the optical properties of the gemstone split the beam into two parallel beams. When the two beams of light exit the stone, they each carry a separate visual image, one with each beam. This property makes it easy to tell the difference between diamond, which is singly refractive, and any diamond simulant that is doubly refractive such as white sapphire, topaz, zircon, or quartz. In fact, the double refraction on some of the diamond simulants, specifically natural zircon, can be so pronounced that it can be seen with the naked eye. The double refraction can be seen as a doubling of the pavilion facet edges when looking into the gemstone through the table facet. On gemstones where the double refraction is very slight, it can often be seen through a jeweller's loupe or uncovered through other gemmological instruments such as the polaroscope.

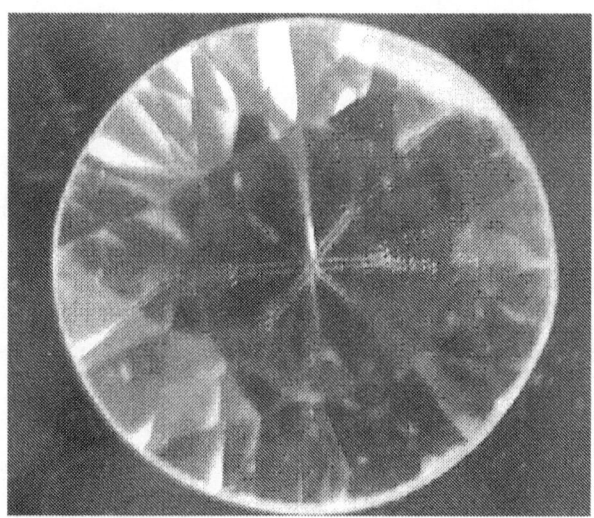

Natural zircon under magnification showing strong double refraction of pavilion facets edges.

Cubic zirconium (CZ) is singly refractive, hence the 'cubic' part of its name, and this is one of the qualities that make CZ a good diamond simulant. In addition, CZ has high dispersion similar to that of diamond. Dispersion is seen as the rainbow flashes or fire that comes from the gemstone. Gemstones that are able to bend a beam of light that passes through them like a prism so that the beam of light is spread out into its spectral colours from red to blue have strong dispersion. Very few gemstones have good dispersion and this is definitely an identifying feature of diamond. In diamond this is often referred to as its fire and this can be seen as small rainbow colour flashes that emerge from a diamond when it is under a strong light source. This property is seen very well when the light source is the sun.

Because CZ has a high dispersion and is also singly refractive, it has been used as a diamond simulant for decades. It has also been used to perpetrate frauds through selling a $4 CZ's as $800.00 or higher priced genuine diamond. I suspect that the supposed diamond ring I stumbled across in that Calgary jewellery store was in fact a CZ, but I did not have further instruments to confirm this at the time.

Both jewellers and consumers have been fooled and defrauded by the criminal use of CZ, however, the jewellery industry developed a simple tool to detect the difference between diamond and simulants. The tool is known as a thermal tester, or the diamond tester, as it is referred to in the industry. It is a simple device that costs no more than $300 and tests the thermal conductivity of a gemstone. Diamond has the greatest thermal conductivity of all gemstones so the instrument is calibrated for diamond. As such, gemstones that do not have the same thermal conductive qualities as diamond are uncovered through the use of this instrument. For many in the jewellery industry, this tool is their primary evaluation tool for determining if the stone they are looking at is a diamond. To them, if the stone looks like a diamond and it tests positive for diamond under the thermal tester, then it is usually considered diamond.

This reliance on the thermal tester rather than gemmological knowledge came back to hurt the industry with the most recent diamond simulant. In the mid 1990's, a new diamond simulant emerged on the market that not only looks like diamond but also tests positive for diamond on the thermal tester. This new simulant has high dispersion like diamond, *appears* singly refractive like diamond, and tests positive for diamond on the thermal tester. This new diamond simulant is known as Moissanite, is created in a lab, and costs about 1/5th the price of diamond. For the purposes of this book and addressing criminal activities, I have called moissanite a diamond simulant, but to be clear on this, Charles and Colvard, the company that creates Moissanite, markets Moissanite as a gemstone in its own right and not as a diamond simulant. This company goes to great lengths to ensure their gemstones are not used in frauds by providing information to the public, industry, and law enforcement about how to detect it from diamond. In this respect alone moissanite stands out because it is proudly being marketed as a created (man-made) gemstone on its own and not as a possible replacement for diamond or imitation. However, since moissanite has been on the market, it has been used in frauds across Canada. When it was first introduced on the market, some jewellers were defrauded and gemmologists were fooled by the new simulant.[141] Since then the jewellery industry has developed a tool that tests for moissanite and as such has eliminated some of the potential for fraud.

When describing the qualities of moissanite that are similar to diamond I said it *appears* singly refractive. I say this because to most people, even those experienced at examining diamonds, moissanite appears singly refractive when the stone is viewed through the table facet. However, when examining moissanite under 10x magnification through crown facets other than the table facet, moissanite appears doubly refractive and is instantly detected as not being diamond. Regardless, most of the general public and law enforcement are unaware of moissanite and many in the jewellery industry have little knowledge. They can still be fooled if they don't know what they are looking for. I was informed recently that the manager of a jewellery store was telling his customers that the world supply of moissanite is mined from the ground in Russia. In fact, moissanite is regularly confused as being a synthetic diamond as is evident in the Criminal Intelligence Service Alberta annual bulletin available on-line through the Alberta Justice web site at www.cisalberta.ca. On page 7 of the 2000–2001 Annual Report, this document says, "synthetic diamond known as 'moissanite' is being smuggled into Canada". This is somewhat problematic because synthetic diamond is not moissanite and moisanite is not synthetic diamond. They are produced differently and handled differently and would be exploited by criminals differently. So the question then is what is being smuggled? It could be that the source of the intelligence doesn't know the difference between moissanite and synthetic diamond or perhaps the information collector got the information mixed up. Whatever the case, like so many people fooled by diamond simulants, law enforcement and the industry have much to learn.

This brings us into synthetic diamonds. These are man made diamonds, chemically identical to their naturally created cousins, but these man-made stones are made in a lab. Man has been trying to make commercially viable diamonds since the 1950's and General Electric Corporation was a pioneer at this work. It is only fitting that man would try to create diamond as most every other gemstone has also been created in labs. These include sapphire, emerald, garnet, amethyst, spinel, and others. The problem at the onset was it was first too difficult to make large diamond crystals and then when they could make large diamond crystals, it was too costly to produce them. Recent advances and new techniques are proving that man-made diamonds of top gem quality can be produced in the lab and can be done so profitably. The problem for the industry and the public is that it is very difficult, even for gemmologists to determine the latest man-made diamonds from the natural diamonds. Gemmologists require new expensive instruments to make these determinations and at present few gemmologists are obtaining these instruments. Even the top labs have been fooled by these new man-made diamonds. There is also the issue of nomenclature. Some in the industry would like to be able to call their man-made or synthetic diamonds, "cultured diamonds". Like all man-made gemstones, disclosure of this fact to the consumer should be a priority. However, the potential for criminal use of man made diamonds is enormous with little chance of repercussions to the criminals.

8

Diamond as a Currency

A senior executive at the production end of the diamond industry once told me "they don't sell diamonds, they sell cash". For business at the top end of the industry this is the case, so much so that before a company will send out its rough diamonds they make sure the buyer has sent them the money first. Literature also speaks of diamonds as a monetary instrument.[142] I don't have to discuss the relative ease of selling diamonds in North America, but for criminals operating here, this is an excellent way to literally store money. Diamonds are money. They are just a different kind of money with different exchange centers than the banks that we are used to in using cash. Criminals will acquire their wealth in diamonds through both legal and illegal activities from points within the jewellery cycle. Jewellery stores, diamond dealers, pawn shops, auctions, the Internet, other criminals, and the public are the exchange centers[143].

The question then is as a monetary instrument, how much is diamond worth? It's not like gold where it has a value per ounce against a countries currency. How can the value of diamond be established as a cash-like instrument? I have spent a lot of time in pawn shops, second hand stores, and with jewellers to establish the pricing schemes they use in purchasing diamonds off the street. There are several factors that can affect the cash value of diamond. Variables could include but are not limited to the country, currency stability, the quality of the diamond, and a person's knowledge of diamonds. However, for North America, there are really only three main factors that impact on the cash value of diamond or its monetary value. The first is the demand.

This is very much a function of where a person lives. There is a very good chance that if someone lives in small town U.S.A. or Canada, that he would not be able to find anyone to buy his diamond from him, at least at a fair market value. Someone may be willing to buy the diamond from him for pennies on the dollar but what use would anyone in a small town have for the diamond aside for personal jewellery? The answer is, very little, and if he was to buy jewellery, it may not be the 'used' piece that someone was selling. When it is not for personal adornment, people buy diamonds from other people because they can make money on reselling the stones to someone else. It maybe that they operate a jewellery store or supply the jewellery industry, but usually diamonds are not bought for the accumulation of wealth unless they would have the opportunity to resell them. That opportunity is somewhat diminished for people living in small rural settings.

In the largest urban centers the demand for diamonds is very high, especially diamonds that can be acquired at prices below what a buyer would normally pay for the stone. The more jewellery stores and auction houses around, the more buyers that are available and as a result the greater demand for the diamond. These conditions make it easier to sell the diamonds someone has. This idea of purchasing a diamond below the price that a buyer would normally pay for a diamond is what fuels this demand and leads us into the next point.

This second point is where someone is positioned on the mine to market chain. This determines what the value of a diamond is to that person on that point of the mine to market chain and therefore what he would expect to receive/pay for the diamond at this point. At the very end of the mine to market chain is the public. They pay the most for diamonds and for them the value of diamonds is the highest. For instance, they may be very happy to purchase a 0.50 carat diamond for $2000 because this may be the typical retail price for such a stone. The next people down on the mine to market chain are jewellers. The jeweller usually buys his diamonds at a slight discount to wholesale price. To jewellers, diamonds are worth much less than to the general public because jewellers can buy diamonds for a lot less. The diamond wholesaler pays even less than a jeweller for diamonds and as such a diamond is worth even less to the wholesaler. The diamond dealers, those who distribute the stones from the diamond cutters, get the finished diamonds at the lowest price and therefore diamonds are worth the least to them. The price of diamonds at the diamond dealer end is the lowest price for finished diamonds and as such it is the lowest cash value that could be expected for a diamond. This doesn't mean that a person could sell diamonds for less than this. It is recognized that people sell their diamonds for much less than they should through selling to pawnshops and second hand stores.[144]

To bring this into perspective, where the public would normally be happy to purchase a 0.50 carat diamond at $2,000, they would be even happier to get it for $1,500. Likewise, the diamond dealers who may normally buy the stone for lets say $600 would be happy to get the same diamond for $500. This also illustrates what the bottom dollar is for a diamond, or what you should expect to get for the stone. The diamond dealers get the diamonds at the lowest prices so if they are willing to buy it then that is likely the bottom dollar. People placed closer to the consumer on the mine to market chain would be happy to get the stone at the price of $500 because they in turn are going to make an even greater profit on that stone. For instance, the jeweller might normally buy the stone from his suppliers for $800, but if he can get it for $500, then his profit margins are significantly higher on that stone. Going further, the average person would be ecstatic to get the diamond they would normally have paid $2,000 for, at the price of $500. The point here is that the value of a diamond is relative to what a person can normally buy it for.

The high cash street value for a diamond, a good price someone should be able to get for a diamond they are selling, should be just slightly below what the diamond dealers pay for diamonds. This is because at this price the diamond dealers can make a profit on the stone and anyone that falls down the chain from them would also make a profit on the stone. I can't say exactly what diamond dealers pay for their diamonds however some diamond experts predict that the bottom dollar for diamond, the price that the diamond dealer pays for stones is approximately 40–45% discount to the wholesale value.[145] The high street value in terms of what the diamond is worth is approximately the price in which it enters the finished diamonds market. This is the high cash street value of diamond as a cash-like instrument if someone needed to exchange the diamond for cash, or convert wealth they had stored in diamonds into cash.

The low cash street level or bottom dollar for diamond is much lower than the high cash street value. This is the value that a criminal who stole the diamond will likely exchange it or sell it for. Criminals who look for immediate exchange through a fence or pawn shop can expect to get about 20–40% of the wholesale price for the diamond in cash or value in product exchanged for the diamonds.

The third factor is the pricing standards like Rapaport, The Guide, and market price. Rapaport pricing represents the high cash wholesale price for diamond of a given quality determined through diamonds sold at the wholesale level in the United States. The Guide is very similar in that it takes into account several factors including wholesale pricing and market pricing to establish a wholesale price for diamonds. The last point is the true market price, which is what it is that diamonds are actually selling for in stores, above established wholesale pricing standards like 'The Guide' and 'Rapaport'. These pricing standards for diamond provide a baseline from which two interested parties, a buyer and seller, can establish a price for the diamond. Without these guides, the value of diamond would be certainly more volatile than it is.

So what is the bottom dollar for a diamond, how much is a diamond worth on the street and how much would someone get for his stone? This would be the price that a person could expect to get for the diamond if he had to sell it quickly and he had a willing buyer, like a jeweller. This is the high cash street value of diamond. I already indicated that one industry expert predicts the diamond would be worth about 40–45% discount to the wholesale guides. For instance, if a person had a diamond with a wholesale value of $1,000, then you could expect to get $600 for the diamond. There are other sources that help confirm these predictions. CBC television did an investigative report on the frauds being perpetrated through the use of moissanite.[146] This expose aired in the late 1990's on a show called Market Place. They had a journalist with a hidden camera and microphone take a ring with a moissanite gemstone around to several jewellery stores pretending to want to sell the ring. The point of the show was to see if anyone would notice that the stone wasn't diamond but rather moissanite, and to illustrate how vulnerable jewellers and consumers are. As it turns out, many of those who offered to purchase the moissanite ring believed it was diamond and Market Place proved this point that both jewellers and the consumer were vulnerable to frauds. What I found particularly interesting about this investigative report was not that they had uncovered the potential for fraud through the use of moissanite, but rather how much the jewellers were prepared to pay for the supposed 'diamond'. The moissanite ring that they used in the show was reported to have a retail value of about $6,000 if it were diamond of similar quality. If using a typical retail mark up of 100%, then one could expect that if the ring they were using was in fact diamond it would have a wholesale value of approximately $3,000. The jewellers that were approached in the show were prepared to pay between $1300 and $2500 for the ring believing it was diamond. This means that they were offering between 43% and 83% of the diamonds expected wholesale value. These numbers certainly support that 60% of the wholesale price (or 40% discount to wholesale) is a high cash street value for quick disposal of a diamond. My personal experience in selling diamonds to jewellers also supports a high cash street value for diamond at about 50–60% of the wholesale value.

Here are a few final words on establishing the street value for diamond. As mentioned, there are so many variables and this number could easily change city-to-city or country-to-country, so these figures are by no means fixed. The amount of money people actually get for a diamond will also be a function of the licit-illicit nature of product they are selling, their knowledge of diamond values, and how quickly they want the money or to dispose of the product. However, the figures represented here are a good baseline for estimating what the high and low cash street values of diamond are.

For criminals who would choose to use diamond for criminal financing and money laundering it could be expected that this would be the high cash street value of the stones in which they deal. Again, if criminals were to purchase diamonds at wholesale pricing, which is very possible especially on bulk sales, they could accumulate a massive storage of cash-like instruments that could be carried on them at all time. Then when they need cash, they can quite easily sell the diamonds at a 20%–30% loss. Based on some publications this is within acceptable losses for laundering money[147]. Coincidently selling them at this price, even factoring in the acceptable money laundering losses, would still mean the criminals could sell the diamonds at the high cash street value or even higher.

The universality and availability of diamond, the demand, the number of venues to dispose of diamond, and the pricing schemes set up make diamond ideal for use as a cash-like commodity. Criminals already know this and are buying and storing diamonds as an alternative to cash in the banks and using diamonds, like cash, to pay for contraband.

9

The Real Threat:
To Industry and Public

The last four chapters have discussed the criminal desire for diamonds and some methods and opportunities in which they can be incorporated into criminal activity. Although from reading this information it may appear that the jewellery industry is rife with criminal activity and this is just not the case. The jewellery industry in general employs very honest people, yet, like any industry or profession, there are those who choose a less than honest path. There are career criminals as well. Like the blood diamond issue, it is not the diamonds that are the problem, it is the people who use them. That said, this industry is wrestling with several issues that are exploitable in a criminal manner. These exploitable opportunities cast a dubious shadow on what is usually thought of as a brilliant business and provide a catalyst for criminal activity. Inflated appraisals, over-grading, simulants, and synthetics are a few practices that present opportunity for dishonest merchants and criminals. These are issues that the industry is working on through public awareness campaigns. In fact, there are hundreds of web sites that offer information to the public on how to safeguard themselves from the deceptive business practices. These web sites are almost exclusively owned and operated by jewellery industry professionals in an effort to protect the public.

These examples represent a sample of the existing exploitation of the jewellery industry. These are also issues that have been historically overlooked by police and where the lack of industry regulation falls short in protecting the industry and those who operate legitimately in it. In terms of police presence, policy, and resources, this is the road least traveled by law enforcement and in turn, the lane of least resistance for criminals; therefore it is precisely where criminals want to operate. Industry response to guideline and regulatory infractions is limited to expulsion from associations or innocuous actions. The penalties have little effect on the offending people or businesses while the public goes on unaware of the unsavory business practices. Despite a loosely regulated industry and associated criminal opportunities, there is little movement towards tighter controls and this is at a time when people working within the industry face increasing threats.

What is the most serious issue of all is the safety of those who work within the industry and those who are targeted by criminals for their jewellery. The criminal desire for diamonds and diamond jewellery means that the acquisition of these items may happen through several serious criminal actions.

Public Threats

Theft, break and enter, robberies, and home invasions are incidents that bring criminals in direct or indirect contact with the public as criminals attempt to acquire jewellery.

Thefts are common occurrences and with respect to the loss of jewellery this can happen in a number of places. Criminals will steal from lockers at a fitness facility and criminals will enter dressing rooms at a hockey arena a take the personal effects found in the pockets of the players clothes. Thefts occur regularly at retail jewellery outlets. The theft can be a result of the opportunity to steal presenting itself to the thief or a calculated, pre-meditated decision to steal.

In Canada, there is little judicial deterrent in committing a residential break and enter to steal jewellery. As a method of obtaining diamond jewellery, breaking into residences is lucrative. One source suggests that in Canada while 1 in 50 houses gets broken into in a year, fewer than 20% of residential break and enters are solved and less than 5% of these incidents results in a court conviction.[148] In addition, while jewellery was among the most common items stolen in residential break and enters in the mid 1990's, there is some empirical and anecdotal[149] evidence that a decade later, jewellery and precious metals may be the most sought after items in residential break and enters.

Robberies and home invasions are another matter. These incidents involve an attack on an individual to obtain property he has. A robbery can happen on a street or alley and with the use of a weapon, a criminal will steal the personal jewellery or anything of value from the victim. A home invasion or residential robbery is similar but the criminal or more often criminals, will forcibly enter the victims' residence to steal their jewellery and valuables from them. These crimes are excessively traumatic and amount to approximately 14% of all robberies in the United States and 7% of all the robberies in Canada.[150]

Industry Threats

Beyond the industry woes, criminal use of diamonds, jewellery, and precious metals is on the rise in the United States and the same trend could be inferred in Canada. As mentioned, between 1999 and 2003, jewellery and precious metal was the fastest growing commodity for thefts and is now only second to automobiles by value.[151] With this, the corresponding rising threat level to those working within the diamond and jewellery industry cannot be ignored. As it stands, diamond and jewellery related crime might be more serious and carry greater ramifications to those in the industry than to other stakeholders. First of all, the criminal activity surrounding diamond and jewellery can have a direct and personal impact on those who work in the industry through thefts, robberies, or organized criminal involvement.

Theft—Results in the loss of product, often uninsured, as well as one's livelihood and alone this can be quite traumatic. In addition the associated damage to a person's store or display can keep the operation closed for several days, increasing the loss and further contributing to skyrocketing insurance costs.

Robbery—All the trauma and turmoil of a theft but magnified through adding an element of violence likely including an edged weapon or firearm.

Extortion—A popular method used by gangs and organized crime to take control of a person's life and/or business through setting them up to fall perhaps through an indiscretion or a drug habit. Once established the criminals can blackmail their target into working to their interest, regardless if it means breaking the law, ethical conduct, or proper procedures.

Coercion—Another method used by organized crime, through the use of threats or intimidation, to gain access to one's business, contracts, or to get the jeweller to work on the criminals' behalf.

The seriousness and regularity of these incidents is illustrated in the headlines and bulletins below:

Recent headlines taken from the Jewelers Security Alliance web site.[152]

ROBBER POINTS HANDGUN AT OWNER
Rochester, MI—May 2, 2006

An owner was alone in a retail jewelry store when a man came in posing as a customer. When the owner opened a showcase, the suspect produced a handgun, pointed it at the owner's face and began yelling. The owner gave the jewelry merchandise to the suspect. The suspect is described as a white male, 5' 8", 26 to 30 years old, blond hair, black raincoat, black and tan floppy hat, and driving a blue Ford Taurus.

TWO SALESPERSONS ROBBED IN INDIANA
Indianapolis, IN—May 1, 2006

At 4:00 p.m., following a day of sales calls, a diamond salesperson was robbed by two masked men in the parking lot of a Starbucks. As the salesperson left the Starbucks and entered his rental car, the suspects blocked him in with their white Chevrolet Lumina. The suspects displayed weapons, smashed the salesperson's car window, and took his backpack of diamonds.

Decatur, IN—April 17, 2006

At 8:35 p.m., following a day of sales calls, a salesman drove to his hotel and checked in. When he returned to his car, three male suspects, who motioned that they were armed, ordered him to lie on the ground. The suspects took his line and slashed a tire of his car. Police were alerted to look for the suspects' green Chevrolet Malibu, which was located, and the suspects arrested after a chase. The three suspects, aged 28 to 31, were in possession of student visas, two of which were expired. The line was recovered.

JEWELER SHOT GOING TO HIS STORE
San Jose, CA—March 14, 2006

At 8:10 a.m. the owner of a retail jewelry store, who was in a rear parking lot on the way to his store, was shot in the leg by a robber. The robber, who concealed his face, took jewelry merchandise that the owner was carrying. The owner, although limping in pain, began work at the store later that day.

SALESPERSON ROBBED WHILE PUMPING GAS
Ft. Lauderdale, FL—March 2, 2006

When a jeweler from New York stopped to pump gas, three suspects used an SUV to block him inside his car. The suspects smashed his rear window and took a line bag from the back seat.

Recent bulletin taken from the Durham Regional Police web site.[153]

Jewellery Store Robbery

On Sunday March 19, 2006, officers from the Oshawa Community Police Office were called to 4C Gems & Jewellery at the Oshawa Centre on King Street West for a report of a robbery.

Shortly after 1:30 p.m. a lone male entered the store and asked about jewellery in the display case. When the male was handed a ring to examine he unzipped his leather coat, displayed a handgun and threatened the store clerk. The male walked from the store and exited through the southeast doors between the Zellers store and the Bay store.

What this all boils down to is while it is easy to illustrate the criminal use of diamonds and like commodities by criminals, the real seriousness of all this is how they are getting diamonds and who are the victims. In every case that law enforcement deals with specific to diamonds and like commodities, the seriousness of the offence is very much in the use of diamonds. But the essence of these offences, the criminal acquisition of the diamonds, can be a far more serious matter. While the criminal use of diamond is often a financial or property crime, the criminal acquisition of the diamonds can often be a persons' crime. Unfortunately the unidentifiable qualities of diamond make it difficult to link the stones to the more serious persons' crimes. And while the initial criminal acquisition of the diamonds, be it a theft, residential break and enter, or jewellery store robbery may not have been committed by the person who was finally found in possession of them, the fact that their actions supported the initial criminal acquisition of diamonds and resulted in serious personal victimization, has to be made clear to the courts.

There is other criminal activity such as the fraudulent use of cheques or credit cards, which is personally innocuous but certainly impacts the industry. There are also large scale frauds that permeate those at the diamond dealing end of the industry who are vulnerable to fraudsters.[154] Even the Russian government, through the Russian Committee for Precious Gems and Metals was defrauded of millions of dollars through an elaborate fraud scam. The scam involved a company named Golden Ada that held a $100-million contract to clean and polish diamonds. The problem was the company would receive the diamonds, then sell them off and the criminals would pocket the cash for the sale.[155]

The jewellery industry is a prime target for these criminals because the merchandise obtained through the fraudulent actions is of high value and easily converted to cash. In addition, personal robberies, home invasions, and weapons offences are commonplace in criminal activity related to jewellery. The newspapers regularly print stories of vicious attacks that have taken place on a clerk or salesman in the jewellery industry or attacks on persons or homes where criminals have targeted jewellery. In 2004, Jewellers Vigilance Canada fanned out over 100 bulletins regarding separate criminal events of jewellery theft or robbery. Most of these bulletins stemmed from activity in Canada and reflects a very small sampling of jewellery related criminal activity.

The use of diamonds to perpetrate frauds, in scams, criminal enterprises, the issue of blood diamonds, terrorist financing, child labour, and other humanitarian questions are exerting downward pressure on the diamond industry. With the criminal potential and use of diamond on the rise, the potential for their activities to tarnish this industry are even greater. The diamond and jewellery industry is built on consumer confidence and these criminal activities have the potential to erode consumer confidence over time. This is a long-term threat to the vitality of the industry and the livelihood of all those who work in it and some in the industry recognize this.[156]

Industry needs to look at current practices at all levels to truncate the exploit opportunities that may be afforded. The same holds true for law enforcement and both sides need to work together in order to make a difference.

10

Terrorist Use of Diamonds

It would be negligent to examine the criminal use of diamonds without touching on the terrorist use of diamonds. There has been a lot written about this subject primarily focused on the use of rough diamonds by terrorist. Douglas Farah is most noted for his books on this topic, plus articles in the Washinton Post and he makes a very compelling argument.[157] However compelling the information provided by Farah and others are, law enforcement has had difficulty substantiating the terrorist use of diamonds. Yet, to think that there is an unlikely possibility of terrorist involvement in the Canadian diamond industry would be naive. This is because terrorist use the same financing mechanisms as organized criminals and organized criminal involvement in the Canadian diamond industry is assured. Then again a lot of this is really a matter of perspective, both on defining who is a terrorist and what exactly is the Canadian diamond industry. Some academics suggest that terrorism is really part of a continuum from basic criminal activity to sophisticated criminals and organized criminals to international criminals and on through to terrorists. In this respect you can't just draw the line that terrorist are all Osama Bin laden clones. It could be argued that when the Rock machine and the Hells Angels were blowing each other up along with any civilians that got in their way during their Montreal turf wars that this was nothing short of domestic terrorism. The same holds true for actions of the F.L.Q (Quebec Liberation Front) in the 70's or the actions of the A.S.A.L.A (Armenian Secret Army for the Liberation of Armenia) operating in Canada in the 1980's[158]. Then again, how far apart in terms of an act of terrorism is what Timothy McVeh did in Oklahoma vs other contemporary acts of terrorism. The objectives of these varied groups or individuals may be different than religious fundamentalism but in every case the actions are extreme.

As mentioned, defining the Canadian diamond industry is also important as a matter of perspective. Often people define the Canadian diamond industry as simply the diamond production or the mining of diamonds. However the Canadian diamond industry is much broader and more complex than simply mining diamonds. This is because the Canadian diamond industry is more than just the production of rough diamonds. It is also the cutting and polishing of these diamonds, some of which is done in less than secure parts of the globe. It is the dealing and retailing of these Canadian diamonds and other diamonds in Canada, because any diamonds, Canadian or otherwise, that are dealt with through the Canadian diamond and jewellery industry are part of the Canadian diamond industry. By comparison, the U.S. diamond industry is very large, inclusive of the New York City Diamond Dealers Club, diamond wholesalers, manufacturers and retailers and personal consumption of diamond jewellery. Add to this that the U.S. does not have any commercial diamond production. Does this then mean they have no diamond industry? Nothing could be further from the truth, because they have a large and vibrant diamond industry. Likewise, just because Canada started producing diamonds in 1998 doesn't mean that Canada didn't have a diamond industry prior to that date. In fact, the Canadian diamond industries history began in the late 1800's. Even though Canada has the world's third highest production of diamonds, a country does not have to have diamond mines to have diamond crimes. Diamond related criminal activity would still be a big issue in Canada even if there were no diamond production. The diamond production just adds to the exploit possibilities. Consider that the largest diamond dealing centre in the world, where nearly 70% of all rough diamonds pass and 50% of all finished diamonds move through, is Belgium, yet they also do not produce any diamonds.[159] If a person defines the diamond industry broadly and defines terrorism broadly, then the possibility of terrorists exploiting diamonds or potential involvement in the Canadian diamond industry cannot be ignored. The same can be said for the diamond industry in the United States and to their credit the U.S. Government is diligently exploring the possibility of terrorist use of diamonds and like commodities through senate committees.[160] However, there are many things that separate the Canadian and United States industries including: the production of diamonds, emerging sub-industry growth, stratification at the production end, and the lack of legal controls, and registration and licensing on the industry and commodity.

Gemstones in general are likely exploit targets for terrorists. Some suggest the Taliban used lapis lazuli (a semi-precious gemstone) from Afghanistan to generate financing.[161] Other reports suggest Al Quaida has been involved in the rough diamond market in Africa for years. Beyond diamonds, it must be noted that valuation of valuable non-diamond, precious gemstones such as ruby, sapphire, and tanzanite is significantly more subjective and could present an even greater exploit potential. Further, countries such as Columbia, Afghanistan, Myanmar, and Sri Lanka are producers of some of the most valuable coloured gemstones as well as home to terrorist organizations, cells, or military regimes. The plausibility of financing these organizations through gemstones, by terrorist organizations with links to Canada, cannot be ignored. At present, what reduces the potential use of these other precious gemstones is a much smaller demand and associated smaller market for resale and therefore a less likelihood of their use.

This limited market is not the case for diamond. It is the most widely sold gemstone in the world and statistically North America consumes about half of the world's annual diamond jewellery production. Remember, this 'half' amounts to approximately $30 billion (U.S.) worth of diamond jewellery being sold to only 5–6% of the world population. This means prolific sales of diamond with correspondingly large amounts of jewellery stores, diamond dealers, and whole-salers to provide diamonds. The criminal or terrorist opportunity to use diamonds in financing efforts are as good in North America as anywhere else in the world as a result of this availability and ease of diamond acquisition as well as the increased ability to return these diamonds to market when cash is desired through an abundance of insertion points.

There is also potential damage to the industry through organized criminal or terrorist involvement in the Canadian diamond industry. There could be an impact at the national or global level that would tarnish the Canadian industry and the nations reputation and damage national and international sales. As an example, what would be the impact to diamond exploration companies and junior miners if a Bre-X style securities fraud was exposed in the Canadian diamond industry? What about a case of a diamond switching scheme related to Canadian diamonds, and what impact would that have on the diamond industry as a whole? What if the switched diamonds are 'conflict diamonds'? How would this impact the confidence in Canadian diamond branding, as well as all businesses downwind from the criminal 'ground zero'? Imagine what would happen if the industry at some intermediate point was connected, even indirectly, to terrorist financing or actions. One could expect serious consequences, especially if talking specifically about Canadian diamonds. The potential exists for a vilification of Canadian diamonds, a boycott by the public or industry, or even a U.S. government trade action against the Canadian industry. This could be disastrous, especially with the United States market consuming nearly half of the world's annual diamond jewellery production. One need look no further than the 1998 U.S. embassy bombings in Tanzania and Kenya, the connection of this event to terrorist financing through the gemstone tanzanite, and the subsequent U.S. boycott of the gemstone. What is of great interest in this case is that the U.S. boycott of this gemstone was largely the action of the U.S. jewellery industries and not a government sanction.[162] In addition and depending on the source there may in fact have been no connection between tanzanite and terrorist financing of this event, yet the consequences to the sales of tanzanite were catastrophic. More recently with Myanmar and its' repressive military regime, there has been movement within the U.S. jewelry industry towards a boycott of gemstones from Myanmar.

If there is one thing learned from 9/11, it is to expect the unexpected. This is by no means fear-mongering or an impossibility. The present size, potential growth, and stratification of the Canadian diamond industry, the global reach of this industry, the North American criminal desire for jewellery, diamonds, and gemstones, and the international criminal involvement in the industry all contribute to the policing challenges that law enforcement face. With this, anticipating the criminal and terrorist possibilities at all levels of the diamond and jewellery industry is the responsibility of law enforcement. This requires police agencies and the diamond and jewellery industry to work proactively to ensure the safety of North Americans and the integrity of our industries.

11

Final Words

Oddly enough Chapter 11 is synonymous with bankruptcy in the United States and while I am not suggesting doom and gloom for the diamond and jewellery industry, I believe that the industry is facing several hurdles now and in the future. I don't have any catchy words of wisdom, simplistic or complex solutions to the problem of criminals using diamonds but what I have observed is that generally speaking, there is little being done by those who can do something about this crime. Again, generally there is a lack of government and law enforcement involvement in specifically and effectively addressing these crimes. There is also a degree of apathy among some within the jewellery industry to acknowledge the problems and then to do something about them. Yet it is likely the jewellery industry that has the most to loose in this battle. In addition, the general public is nearly ignorant of these issues. Perhaps if they were not, their voices and pressure would have an affect on those who can do something about this.

The industry has been built on consumer confidence and this is nurtured and maintained in several ways. This includes but is not limited to a trust in what is being purchased and in the integrity of the industry and those who work in it. It is also a function of diamonds and what they stand for.

We know that society has revered diamonds as precious and sentimental items worthy of gifts of jewellery in the most precious moments of our lives. This has been a fundament use for diamonds in North American society. To a lesser extent diamonds have been used for investment. The use of gem diamonds is now changing in contemporary society and expanding as an instrument for investment. There is at least one major investment corporation in the United States that may soon offer up diamonds as an investment. This change in the way diamonds are used also affects the perception of diamonds, good or bad, and the sentimental value they have historically held with consumers. In maintaining the diamond industries sterling image this presents one more challenge in the industries drive to hold consumer confidence vis-à-vis disclosure, transparency, the

Kimberley Process. However, what permeates all of this is the criminal use of diamonds and their involvement in the diamond and jewellery industry. This may be the greatest threat to consumer confidence that the industry must content with. The use of diamonds and like commodities by criminals has put the spot light on the diamonds and jewellery industry and has led to new laws, regulations and scrutiny that did not exist ten years ago. This impacts the industry in terms of optics and how it is perceived and in the way business is done with the consumer. The latest laws engage issues and criminal activities that are damaging to the diamond and jewellery industry on several levels and present new and/or enormous policing challenges. While there are various criminal uses for diamonds, the contemporary issues of blood diamonds, money laundering, and counterfeit products could be expected to have the greatest impact.

Blood Diamonds

The work of non-government organizations exposed the horrific story of how rough diamonds have been traded to illegal arms dealers for weapons. These weapons have been used in civil conflicts in West African countries and have resulted in widespread death and atrocities. In several respects this story is not new; despite the fact it was brought to the worlds attention in the late 1990's. The use of diamonds to pay for war, protection or to cover war costs dates back several centuries and can be found in the great pedigree of the 108 carat, Kohinoor diamond. In this respect diamonds, like gold, oil or any other valuable commodity could be used to fund a war. It just so happens that diamonds are a cash-like instrument and can be traded directly for weapons or used as an alternative to cash. Other commodities typically have to be sold in their respective markets for cash and then the cash used to buy the weapons.

It has been about a decade since the height of the blood diamond era and the industry and most diamond producing countries have taken extensive measures to curb the blood diamond problem. In the late 1990's it was estimated that the amount of blood diamonds being traded was 4% of the world gem diamond supply. Today, with reductions in West African conflicts and the implementation of the Kimberley Process, the trade of blood diamonds has been reduced. In response, the industry has undertaken awareness programs and implemented chains of warranties down-stream from the rough diamond industry to authenticate diamonds as being conflict free. Governments have followed Kimberley Process protocols by providing certification programs for rough diamonds and enacting legislation to back the process. Despite this the Kimberley process has not been without its problems and smuggling and illegal trade continues. This on-going criminal trade of blood diamonds negatively affects the consumers' perception of the industry with each and every new event and news release.

However, like the Kohinoor that is presently mounted in the Queen Mothers crown, the diamond is not the problem. Similarly, the blood diamond issue must be considered in context. The West African wars and accompanied atrocities could be fuelled by any number of commodities, but diamonds happen to be a local currency in this part of the world. The problem ultimately rest at the feet of those people who will use diamonds or whatever "currency" for criminal purposes or to fund a war. Equally at fault are those who participate by taking diamonds in trade for arms or to fund their criminal empires and those who continue to conduct business without committing to the KP protocols. Discontent that may be generated by the criminal use of diamonds in this manner should be directed not at diamonds or the diamond industry in general, but at those criminals and warring factions and industry participants who have exploited these diamonds. While industry, government, and law enforcement the world over all have a stake in this issue, the industry takes the gold medal for its efforts in consumer awareness and reducing the blood diamond trade.

Money Laundering and The Associated Crimes

Blood diamonds connected to tragic events in West Africa taint the diamond industries lustrous image but has not significantly impacted the consumer or their desire to purchase diamonds. It has more likely resulted in a shift of consumer demand; one of which the Canadian diamond industry has likely benefited. On the other hand, anti-money laundering legislation requires jewellers to pry personal information from the consumer and reduces the experience of purchasing jewellery (a.k.a romancing the stone) to that of a bank loan. This hits the consumer directly at point of sale.

We know that criminals have utilized diamonds to fund their activities and are increasingly turning to this commodity for the purpose of money laundering. Yet, in many respects coloured gemstones are as good as or superior to diamonds especially for facilitating money laundering because of the unfixed values they hold. As such, anti-money laundering legislation cannot simply be committed to diamonds and the popular precious gemstones (a.k.a Noble gemstones) such as ruby, sapphire, emerald etc. Legislation of such a limited scope would be like having a national drug strategy that focuses on marihuana and cocaine but leaves out crystal meth or heroin.

This legislation must be comprehensive and include all common gemstones both organic and inorganic. What is important to know is that if the anti-money laundering legislation component relative to diamonds and gemstones simply focuses on diamonds and six or eight other gemstones, then most certainly criminals will turn to those gemstones that are not subject to the legislation. On this point, it is interesting to note that garnet, amethyst, aquamarine, diamond, emerald, alexandrite, ruby, peridot, sapphire, pearl, citrine/topaz, tanzanite/zircon are the birthstones from January to December inclusive. It is obvious to see how widely accepted and traded they all are by examining what is for sale in your local jewellery store. Yet only a handful of these stones are included in money laundering legislations around the world. It is not uncommon for andradite or tsavorite garnet to be priced well beyond that of tanzanite, sapphire, ruby or emerald—it just depends on the quality. Similarly, Topaz is one of the biggest selling gemstones in North America by volume but it is not considered to be a threat. Then there are pearls. A single sea pearl can easily price out over $2000 each and high-end 16" strands of south sea pearls can wholesale over $100,000. It would be nice to see these items show up in legislation as well. Then again legislations also need to define what a particular gemstone is. For instance, in some parts of the world emerald is only considered emerald if it gets its colour from chromium. Otherwise it is just green beryl. In other parts of the world it doesn't matter if there is chromium or not for the stone to be called emerald. Lack of international standards or even a definition can present a problem here. Similar identification issues exist with pink sapphire and ruby. Then again what about man-made diamonds or treated diamonds? Could they by way of definition be left out of legislation? While this book has been written about diamonds, in reality nearly any other gemstone could be utilized criminally in a similar manner as diamonds. Further, the criminal use of precious metals for money laundering or otherwise is equally problematic yet all together different than diamonds.

Consider that police recover about $40 million of the $1 billion worth of diamonds and like commodities stolen every year in the United States. This means that about $960 million worth of diamonds, gemstones, jewellery and precious metals are being held or utilized by criminal to further other criminal activities or are laundered by reinserting them back into the legitimate market—every year. This doesn't even consider the amounts stolen and used by criminals in Asia and Europe and the rest of the world or proceeds of crime used to outright purchase diamonds and diamond jewellery. By comparison, the global blood diamond trade, estimated at 4% of the $7 billion international rough diamond production at its peak, translates loosely into about $280 million. This truth fundamentally supports the creation and application of money laundering legislation directed at the precious gemstones and metals sector in North America than in any other continent. Also, this should spur the industry and law enforcement into equally stringent measures to eliminate the criminal use of diamonds and like commodities. This is especially so because this issue is very much home grown, North American, and not the realm of some far off West African country. This affects us daily with every diamond and jewellery specific burglary, robbery, or theft and with the purchase of recycled product that has been tainted by crime. The criminal use of diamonds and like commodities especially in respect of money laundering is definitely a growth industry.

Copyright Infringement

This issue within the jewellery industry goes beyond jewellery itself and includes other products such as counterfeit diamond grading certificates and the branding of diamonds. I have seen counterfeit jewellery in so many stores now that I would describe it as an epidemic within the jewellery industry. I can say this with some certainty for Canada and limited certainty for the United States (I have not traveled as extensively in the U.S.). What is surprising is how many brick and mortar jewellery stores actually carry these products. In my travels and in talking with jewellers who sell counterfeit jewellery, most of them know the products they are selling are counterfeit. While many of them are probably unaware of the particulars of legislation that prohibits the sale of counterfeit products, it is clear they know they are doing something wrong. However, I suspect that most of those who sell counterfeit jewellery do not realize the potential severity of their actions criminally or civilly. They certainly do not understand the negative impact they have on the jewellery industry and if they do, they just don't care.

To be clear, the majority of retailers I have seen do not carry any counterfeit goods and there are some stores that may have just a few pieces in stock and while I have seen others that are nearly completely filled with counterfeit jewellery pieces. Further, this is nearly always gold or silver jewellery, and not diamonds. From the perspective of a consumer and someone who sells jewellery, this issue casts a dark shadow over the jewellery industry. As such all products sold in these stores, including diamonds, are tainted directly or indirectly by this.

There are several negative forces all working in parallel in the production and sale of counterfeit jewellery. On one level the production and sale of counterfeit jewellery may be supporting the financial gains of organized criminals and beyond that the jewellery stores involved are committing a crime by selling the counterfeit jewellery. The damage continues with the loss of revenue to those who hold copyright on the products being counterfeited. From a consumer perspective and perhaps the biggest problem for the industry is that the consumer is being defrauded. Regardless, the savvy consumer who recognizes counterfeit jewellery in a retail location has to wonder what other criminal activity could exist there or what industry rules are being bent or broken at that store. In my experience I have found that copyright infringement has gone hand in hand with some other criminal or regulatory infractions. Once the consumer has uncovered the fraud there could be several ramifications including, civil litigation and if they get the police involved, a criminal prosecution. Perhaps a civil litigation may be avoided if the consumer is remunerated, refunded or a settlement is reached. However, the involvement of the police in these types of investigations is serious on two fronts. While criminal charges and the resulting criminal penalty may involve a forfeiture of any counterfeit product, a criminal record, monetary fine and even jail time, this may just be the beginning. In any criminal investigation, the brand name holder or those who otherwise hold the copyright of the items counterfeited will authenticate the counterfeit product seized. The police will involve these companies to gather evidence for their criminal case and in doing so these companies are made well aware of the counterfeit infraction. As mentioned before, these companies that hold the copyright are also victims and they will often engage in a law-suit against the offending company. These companies spend millions of dollars on building their brand name and protecting their copyright and protect their interests in criminal and civil court. Financially, the impact of the civil proceeding on the offending company may be far more severe than anything handed down by the criminal courts.

One thing is certain, regardless if they disclose that the product is a knock off, if the design, name, logo or trademark is an unlicensed reproduction, then it is all but certainly a crime to sell the product. Tackling this is a huge resource demand on law enforcement because the larger framework of this issue encompasses clothing, electronics, music and nearly every other consumer product. However, the issue of counterfeit jewellery could be dealt with to some degree by the industry through self-regulation, association standards and using other industry watchdogs such as the Better Business Bureau. Further, industry and law enforcement working together on this could significantly disrupt the North American counterfeit jewellery problem.

When all is said and done the diamond and jewellery industry is under the gun as a function of criminal use of diamonds and like commodities. The diamond and jewellery industry, law enforcement, and government all have a stake and specific role in tackling and eliminating the criminal use of diamonds. Now more than ever these industries need the support of law enforcement to stamp out the criminal exploitation of diamonds and jewellery. This means acknowl-edging the criminal use of diamonds and like commodities by law enforcement as a specific type of crime and finding resources to deal with this problem. It also means consulting with the industry to find out how best law enforcement can be of service and immersing both practically and academically in the jewellery indus-try. These experiences can be obtained through one of the several schools of gemology listed in Appendix B, through attending events such as the Gemboree in Bancroft, Ontario or the Canadian Gemmological Association annual confer-ence, and simply getting to know your local jeweller. Law enforcement also needs to educate the industry and public through providing awareness of the criminal use of these commodities and crime prevention measures and to follow up with industry specific crime prevention programs. Likewise the diamond and jewellery industry needs to step up and take back their industry from criminals who are eroding the legitimate market at all levels. This means reinforcing everything that is positive within the industry, supporting legitimacy, honesty and professional business practices at all levels and amending where consumers are being exploited. This is very much about being united, self-regulation, transparency, and disclosure and it goes along way towards building trust and reinforcing legit-imacy—otherwise increasing consumer confidence. This is certainly not beyond industry grasp as the industry has already done excellent work with the blood dia-mond issue in this regard. The third element here is the government. Govern-ment needs to help by creating meaningful effective legislation and regulation for this industry that marginalizes the exploit potential and provides law enforce-ment with tools to deal with criminals. This also includes providing industry sup-port in implementing legislation and regulations. Using the money laundering legislation for instance: public awareness initiatives and point of sales information (such as pamphlets) could be extremely beneficial to smoothing the transition in how sales are done in the jewellery industry. Such initiatives take the heat off of the retailer in gathering information from a customer. There may be a further added value in providing this support through gaining higher industry compli-ance. The fifth C, "Certified", "Conflict free" or whatever it may be has the potential to bring added value and legitimacy to the diamond industry and with it added consumer confidence. Left to fester, criminality becomes interwoven with diamonds and the diamond industry and the fifth C could become cemented as crime, with all of its negative attributes.

APPENDIX A

Author Contact

Kelly Ross
#140, 240-222 Baseline Road
Sherwood Park, Alberta
Canada T8H 1S8

Email; RossInc@telus.net
Web Site; www.RossInc.ca

APPENDIX B

Information Sources

Gemmological Labs and Training

Harold Weinstein (Canada)
www.hwgem.com

Gemological Institute of America (U.S.A) www.gia.edu/gemtradelab/31494/section_main_page.cfm

International Gemological Institute (Belgium, Canada)
www.igiworldwide.com

European Gemmolgical Laboratory (U.S.A)
www.eglusa.com

American Gemological Society (U.S.A)
www.agslab.com

Canadian Gemmological Association (Canada)
www.canadiangemmological.com

Canadian Institute of Gemmology (Canada)
www.cigem.ca

Useful Websites

The Canadian Anti-Counterfeiting Network www.CACN.ca
Jewellers Vigilance Canada www.jewellersvigilance.ca
Canadian Jewellers Association www.canadianjewellers.com
Jewelers of America www.jewelers.org
Jewelers Security Alliance www.jewelerssecurity.org

CIBJO (World Jewellery Confederation) www.CIBJO.com
Alberta Jewellery Industry Crime Watch www.AJICW.ca
K Division RCMP Diamond Program www.rcmp-grc.gc.ca/ab/prog_serv/diamond_e.htm

Documents on-line

The Guidelines with Respect to the Sale and Marketing of Diamonds, Coloured Gemstones, and Pearls
www.jewellersvigilance.ca/html/industry_standards.htm

The Voluntary Code of Conduct for Authenticating Canadian Diamonds
www.canadiandiamondcodeofconduct.ca

Deceptive Marketing Practices
www.competitionbureau.gc.ca

The Kimberley Process
www.cbsa-asfc.gc.ca/E/pub/cm/d19-6-4/d19-6-4-e.html

Related Canadian Laws

http://laws.justice.gc.ca/en/BrowseTitle
Weights and Measures Act
Customs Act
Criminal Code
Copyright Act
Competition Act
Export and Import of Rough Diamonds Act
Cultural Property Export and Import Act
Precious Metals Marking Act

Public Events

Gemboree, Bancroft Ontario, August long weekend,
http://www.bancroftdistrict.com
Calgary Rock and Lapidary Club, Annual Show, May, www.crlc.ca
CGA Annual Conference, http://www.canadiangemmological.com

Recommended Books

Diamonds : by M. Hart
Blood from Stone : by D. Farah
Cold Terror : by S. Bell
Confessions of a Master Jewel Thief : by B. Mason
Diamond Industry Strategies to Combat Money Laundering and the Financing of Terror : by C. Even—Zohar
Nature of Diamonds : by G. Harlow
Gemmology : by P. Read

Endnotes

1. Harlow G, 1998, *The Nature of Diamonds*, American Museum of Natural History, Cambridge University press, Cambridge, United Kingdom

2. Harlow G, 1998, *The Nature of Diamonds*, American Museum of Natural History, Cambridge University press, Cambridge, United Kingdom

3. Harlow G, 1998, *The Nature of Diamonds*, American Museum of Natural History, Cambridge University press, Cambridge, United Kingdom

4. Hart, M 2001, *Diamonds*, Penguin Books, Toronto, Canada

5. Harlow G, 1998, *The Nature of Diamonds*, American Museum of Natural History, Cambridge University press, Cambridge, United Kingdom pp.111

6. Wikipedia web site, "The Golden Jubilee", viewed on dec 22, 2004, at <http://en.wikipedia.org/wiki/The_Golden_Jubilee>

7. Arkansas State Park web site, "Notable Diamonds From Arkansas Diamond Site", viewed on December 22, 2004, at <http://www.arkansasstatepark.com/media/cd_article.asp>

8. Arkansas State Park web site, "Notable Diamonds From Arkansas Diamond Site", viewed on December 22, 2004, at <http://www.arkansasstatepark.com/media/cd_article.asp>

9. Ward, F 1993, *Diamonds*, Gem Book Publishers, Bethesda, U.S.A.

10. Arkansas State Park web site, "Notable Diamonds From Arkansas Diamond Site", viewed on December 22, 2004, at <http://www.arkansasstatepark.com/media/cd_article.asp>

11. Gemological Institute of America 2006, retrieved from <http://www.gia.edu/newsroom/4026/image_gallery_list.cfm>

12. Northwest Territories, Energy, Mines and Petroleum Resources 1993, *Diamonds and the Northwest Territories, Canada*, Northwest Territories, Energy, Mines and Petroleum Resources, Government of the Northwest Territories, Yellowknife, Canada

13. Tailby, R 2002, "The Illicit Market in Diamonds", *Trends and Issues in Crime and Criminal Justice, Australian Institute of Criminology*, Canberra, Australia, no. 218.

14. Criminal Intelligence Service Canada 2005, *Annual Report on Organized Crime 2005*, Criminal Intelligence Service Canada, Ottawa, Canada

15. Hart, M 2001, *Diamonds*, Penguin Books, Toronto, Canada

16. Harlow G, 1998, *The Nature of Diamonds*, American Museum of natural History, Cambridge University press, Cambridge, United Kingdom pp.77

17. Harlow G, 1998, *The Nature of Diamonds*, American Museum of natural History, Cambridge University press, Cambridge, United Kingdom.

18. Hart, M 2001, *Diamonds*, Penguin Books, Toronto, Canada pp.161

19. Yiannopoulos, N 2003, "A Guide to Rough Diamond Classification", *Rough Diamond Review* June 2003, Volcano Publishing, Northbridge, Australia

20. Peters, N 1998, *Rough Diamonds—A practical guide*, American Institute of Diamond Cutting, Ft. Lauderdale, Florida, U.S.A

21. Hart, M 2001, *Diamonds*, Penguin Books, Toronto, Canada

22. Ward, F 1993, *Diamonds*, Gem Book Publishers, Bethesda, U.S.A

23. Read, P 1999, *Gemmology—second edition*, Butterworth-Heinemann, Woburn, U.S.A

24. Peters, N 1998, *Rough Diamonds—A practical guide*, American Institute of Diamond Cutting, Ft. Lauderdale, Florida, U.S.A

25. Gemological Institute of America 2006, retrieved from <http://www.gia.edu/newsroom/4026/image_gallery_list.cfm>

26. Cuellar, F 1998, Diamonds For Profit, ISBN 0-9668131-1-1

27. Tiffany & Co. web site, "Tiffany & Co. Celebrates Opening of Laurelton Diamonds Facility in Canada's Northwest Territories", viewed on March 1, 2005 at <http://www.shareholders.com/tiffany>

28. Hart, M 2001, *Diamonds*, Penguin Books, Toronto, Canada

29. Answer $5050.00 (1.01 carat x $5000/ct = $5050.00)

30. Pagel—Theisen, V 1993, *Diamond Grading ABC—Handbook for Diamond Grading*, Rubin and Son bvba, Antwerp, Belgium.

31. Pagel—Theisen, V 1993, *Diamond Grading ABC—Handbook for Diamond Grading*, Rubin and Son bvba, Antwerp, Belgium.

32. Wikimedia Web site 2006, retrieved from <http://commons.wikimedia.org/wiki/Image:Diamond_facets.gif>

33. Gemological Institute of America, retrieved from <http://www.gia.edu/newsroom/4026/image_gallery_list.cfm>

34. Pagel—Theisen, V 1993, *Diamond Grading ABC—Handbook for Diamond Grading*, Rubin and Son bvba, Antwerp, Belgium.

35. Pagel—Theisen, V 1993, *Diamond Grading ABC—Handbook for Diamond Grading*, Rubin and Son bvba, Antwerp, Belgium.

36. Gemological Institute of America, retrieved from <http://www.gia.edu/newsroom/4026/image_gallery_list.cfm>

37. Department of the Treasury 2001, *Financial Crimes Enforcement Network; Anti-Money Laundring Programs for Dealers in Precious Metals, Stones, or Jewels*, Department of The Treasury, Fincen, Washington, U.S.A (31 CFR Part 103).

38. Jewellers Vigilance of Canada website, "Canadian Charged with Smuggling", viewed May 3rd, 2006 at, <www.jewellersvigilance.ca/html/jvc_actionup.thm>

39. Ernst & Young 1997, *Report on the Excise Tax on Jewellery*, Ernst & Young, viewed on July 13th, 2001, at <http://cja.polygon.net/excise>

40. Global Congress on Combating Counterfeiting website, "The First Global Congress on Combating Counterfeiting: Recommendations", viewed May 3rd, 2006 at, <http://www.anti-counterfeitcongress.org>

41. Gemological Institute of America web site, "GIA Alerts Trade as Counterfeit Grading Reports Surface in Antwerp", viewed January 22, 2006 at, <http://www.gia.edu/newsroom>

42. Professional Jeweler web site, "Tiffany & Co. Stops Counterfeiter", viewed January 22, 2006 at, <http://www.professionaljeweler.com>

43. Canadian Guidelines with respect to the Sale and Marketing of Diamonds, Coloured Gemstones and Pearls, Revised Edition 2003, pp. 1, Established by the Jewellers Vigilance of Canada Inc.

44. The Canadian Diamond Code Committee 2006, *The Voluntary Code of Conduct for Authenticating Canadian Diamond Claims*, The Canadian Diamond Code Committee, Toronto, Canada

45. The Canadian Diamond Code Committee 2006, *The Voluntary Code of Conduct for Authenticating Canadian Diamond Claims*, The Canadian Diamond Code Committee, Toronto, Canada, pp.1

46. Hollington, K 2004, *Diamond Geezers: The Inside Story of the Crime of the Millenium*, Michael Omara publishing, ISBN 1843171228

47. U.S. General Accounting Office GAO-04-163, 2003, "Terrorist Financing: U.S. Agencies Should Systematically Assess Terrorists use of Alternate Financing Mechanisms", U.S. General Office of Accounting, Washington D.C, U.S.A

48. Federal Bureau of Investigation 2004, *Crime in the United States 1999–2003, Federal Bureau of Investigation*, Washington D.C., U.S.A

49. Federal Bureau of Investigation 2004, *Crime in the United States 1999–2003, Federal Bureau of Investigation*, Washington D.C., U.S.A

50. Robinson, J. 2003, *The Sink*, McClelland & Stewart Ltd., Toronto, Canada

51. Hart, M 2001, *Diamond*, Penguin Books, Toronto, Canada

52. Even-Zohar, C 2004, *Diamond Industry Strategies to Combat Money laundering and the Financing of Terrorism*, ABN-AMRO, Belgium.

53. Pricescope web site, "Prices of Diamonds Graded by Three Different Laboratories", viewed Oct 17, 2004 at; <http://grading.pricescope.com>

54. Jewelers Circular Keystone wed site, E. Rosen, 1997 "Appraising for Consumer Resale", viewed on March 1st, 2006 at;
<http://www.jckgroup.com>

55. Daily Times web site, "Saddam's Thai gem Spree Hints at Getaway Plan", viewed on May 5th, 2006 at; <http://www.dailytimes.com.pk>

56. Ward, F 1993, *Diamonds*, Gem Book Publishers, USA.

57. Criminal Intelligence Service Canada, "2003 Annual Report on Organized Crime in Canada", Criminal Intelligence Service Canada.

58. Criminal Intelligence Service Alberta, "2001 Annual Report on Organized Crime in Alberta", Criminal Intelligence Service Alberta

59. Robinson, J. 2003, "*The Sink*", McClelland & Stewart Ltd., Toronto, Canada

60. Financial Action Task Force on Money Laundering 1999, *1998–1999 Report on Money Laundering Typologies*, Financial Action Task Force, February 1999, Paris, France.

61. Schneider, S 2004, "*Money Laundering in Canada: An Analysis of RCMP Cases*", Nathanson Centre for the Study of Organized Crime and Corruption, Toronto, Canada

62. Robinson, J. 2003, "*The Sink*", McClelland & Stewart Ltd., Toronto, Canada

63. United Nations web site 2006, "Global Programme Against Money Laundering", United nations Office on Drugs and Crime, viewed on June 6, 2006 at <http://www.unodc.org/unodc/money_laundering.html>

64. Carl Levin United States Senator Web site 2006, "Money laundering Costs All of Us", viewed June 6, 2006 at <http://www.senate.gov/-levin/newsroom/release.cfm?id=210680>

65. Ward, F 1993, *Diamonds*, Gem Book Publishers, USA.

66. Tacy Ltd. Web site 2005, Diamond Trade Cited as Vulnerable to Money Laundering by U.S. State Department Report, Tacy td. Diamond Industry Consultants, viewed May 22, 2006, at <http://www.tacyltd.com/Research_Materials_Full.asp?id=53421>

67. U.S. General Accounting Office GAO-04-163, 2003, "Terrorist Financing: U.S. Agencies Should Systematically Assess Terrorists use of Alternate Financing Mechanisms", U.S. General Office of Accounting, Washington D.C, U.S.A

68. Levy, L Prof. Circa 1980, Antwerp Diamond Bourse, Belgium Antwerp.

69. Callahan, M.S. 1996, "*Insider Secrets to Diamonds Dealing*", Paladin Press, Boulder, Colorado

70. Farah, D 2005, "Terrorist responses to Improved U.S. Financial defenses", before the House Subcommittee on Oversight and Investigations Committee on Financial Services, delivered February 16th, 2005.

71. Federal Bureau of Investigation 2004, *Crime in the United States 1999–2003, Federal Bureau of Investigation*, Washington D.C., U.S.A

72. Read, P.G. 1999, *"Gemmology—second edition"*, Butterworth-Heinemann, Wolburn, MA, U.S.A

73. Scott, M 2004, "Burglary of Single family Houses in Savannah Georgia", United States Department of Justice, Office of Community Oriented Policing Services, Washington.

74. Carmicheal, A 2004, "Organized Crime Growing in Canada—Police Chiefs Ask for Public Help", in *The Telegram*, August 21, 2004, St. John's, NFLD

75. Northwest Territories Resources, Wildlife and Economic Development 2004, "2004 Diamond Industry Report—Diamond Facts", Yellowknife, NWT.

76. Munusamy, R 2002, "Secret Agents Bust Diamond Dealers", Sunday Times newspaper, viewed on June 23, 2002 at, <http://ww.sundaytimes.co.za>

77. Ward, F 1993, *Diamonds*, Gem Book Publishers, USA.

78. Mason, B 2005, *Confessions of a Master Jewel Thief*, Villard Publishing, ISBN 0375760717

79. Shor, R 1997, "Auction Houses vs Luxury Retailers Myth vs Reality", in Jewelers Circular Keystone Magazine January 1997.

80. Canoe web site, "Robinson Charged in Diamond Theft", in the Ottawa Sun viewed February 10th, 2005, at; <http://www.canoe.ca/NewsStand/OttawaSun/News/2004/06/22/508945.html>

81. Nelson D., Collins L., and Gant F., 2002, "The Stolen Property Market in the Australian Capital Territory", Australian Institute of Criminology for the ACT Department of Justice and Community Safety, Australia

82. Weatherburn, D 1998, "The Stolen Goods Market in New South Wales", NSW Bureau of crime Statistics and Research, NSW Attorney Generals Office, November 2, 1998.

83. Federal Bureau of Investigation 2004, *Crime in the United States 1999–2003, Federal Bureau of Investigation*, Washington D.C., U.S.A

84. Cuellar, F 1998, *Diamonds For Profit*, pp. xiii, ISBN 0-9668131-1-1

85. Cuellar, F 1998, *Diamonds For Profit*, pp. xiii, ISBN 0-9668131-1-1

86. Eisen, S 1995, "To Tell or Not to Tell: Should Your Customers Know it's a Used Diamond", viewed on Jewelers Circular Keystone Magazine web site on March 24, 2005 at; <http://www.jckgroup.com>

87. Hicks, D and Sansfacon, D 2000, "Preventing Residential Burglaries and Home Invasions" for the International Centre for the Prevention of Crime, viewed on November 28th, 2004 at; <http://www.crimeprevention-Intl.org>

88. Nelson D., Collins L., and Gant F., 2002, "The Stolen Property Market in the Australian Capital Territory", Australian Institute of Criminology for the ACT Department of Justice and Community Safety, Australia

89. Weatherburn, D 1998, "The Stolen Goods Market in New South Wales", NSW Bureau of Crime Statistics and Research, NSW Attorney Generals Office, November 2, 1998.

90. Mason, B 2005, *Confessions of a Master Jewel Thief*, Villard Publishing, New York, USA.

91. Clarke, R 1999, "Hot Products: understanding, anticipating and reducing demand for stolen goods", Policing and Reducing Crime Unit, Research, Development and Statistics Directorate, London, England.

92. Hicks, D and Sansfacon, D 2000, "Preventing Residential Burglaries and Home Invasions" for the International Centre for the prevention of Crime, viewed on November 28th, 2004 at; <http://www.crimeprevention-Intl.org>

93. Hicks, D and Sansfacon, D 2000, "Preventing Residential Burglaries and Home Invasions" for the International Centre for the prevention of Crime, viewed on November 28th, 2004 at; <http://www.crimeprevention-Intl.org>

94. Stevenson R, and Forsythe L, 1998, "The Stolen Goods Market in New South Wales", NSW Bureau of crime Statistics and Research, NSW Attorney Generals Office, November 2, 1998.

95. Severs L, 2006, "Diamond Miners Eyeing Bigger Carat Crop", in the Business Edge Alberta Business News, 2006 February 16, vol. 6, no. 4, Edmonton, Canada, viewed at <www.businessedge.ca>

96. Hemmingway, J 1992, "Diamond Engagement Ring 1992 Fact Sheet", *Selling Diamonds*, Diamond Promotion Service, USA

97. Hemmingway, J 1992, "Sweet 16 Diamond 1992 Market Fact Sheet", *Selling Diamonds*, Diamond Promotion Service, USA

98. Hicks, D and Sansfacon, D 2000, "Preventing Residential Burglaries and Home Invasions" for the International Centre for the prevention of Crime, viewed on November 28th, 2004 at; <http://www.crimeprevention-Intl.org>

99. CBC Unlocked web site, "Banks, Credit Union tighten debit fraud complaint process", viewed March 17, 2006, at; <http://www.cbcunlocked.com/artman/publish/article_560.shtml>

100. Garcia-Swartz, D & Hahn, R 2006, "*The Move Toward a Cashless Society: A Closer look at Payment Instrument Economics*", in the Review of Network Economics, Volume 5—issue 2, June 2006, AEI Brookings, Washington, DC.

101. Tailby, R 2002, *The Illicit Market in Diamonds*, Trends and Issues in Crime and Criminal Justice No. 218, Australian Institute of Criminology, Canberra, Australia.

102. Tailby, R 2002, *The Illicit Market in Diamonds*, Trends and Issues in Crime and Criminal Justice No. 218, Australian Institute of Criminology, Canberra, Australia.

103. Tailby, R 2002, *The Illicit Market in Diamonds*, Trends and Issues in Crime and Criminal Justice No. 218, Australian Institute of Criminology, Canberra, Australia.

104. Global Witness 2000, *Conflict Diamonds—Possibilities for the Identification, Certification and Control of Diamonds*, Global Witness Ltd., London, UK

105. Tailby, R 2002, *The Illicit Market in Diamonds*, Trends and Issues in Crime and Criminal Justice No. 218, Australian Institute of Criminology, Canberra, Australia.

106. Global Witness 2000, *Conflict Diamonds—Possibilities for the Identification, Certification and Control of Diamonds*, Global Witness Ltd., London, UK

107. Tailby, R 2002, *The Illicit Market in Diamonds*, Trends and Issues in Crime and Criminal Justice No. 218, Australian Institute of Criminology, Canberra, Australia.

108. Even-Zohar, C 2005, *Terror and Diamonds in Washington*, Tacy Ltd. Diamond Industry Consultants, viewed on May 22, 2005 at <http://www.tacyltd.com/Research_Materials_Full.asp?id=54833>

109. Partnership Africa Canada 2006, '*Diamond Controls in Venezuela—100% KP "Non-Compliance"*, in Other Facets Number 22 December 2006, Partnership Africa Canada, Ottawa.

110. Tailby, R 2002, *The Illicit Market in Diamonds*, Trends and Issues in Crime and Criminal Justice No. 218, Australian Institute of Criminology, Canberra, Australia.

111. Tacy Ltd. 2005 web site, *Crackdown on Illicit Diamond Diggers in Angola*, Tacy Ltd. Diamond Industry Consultants, viewed May 22, 2005, at <http://www.tacyltd.com/research_materials_full.asp?id=55472>

112. Blore, S 2006, "The Lost World: Diamond Mining and Smuggling in Venezuela", Occasional Paper #16, Partnership Africa Canada, Ottawa

113. Blore, S 2006, "The Lost World: Diamond Mining and Smuggling in Venezuela", Occasional Paper #16, Partnership Africa Canada, Ottawa

114. Even-Zohar, C 2004, *Diamond Industry Strategies to Combat Money Laundering and the Financing of Terrorism*, ABN-AMRO, Belgium.

115. Kaiser, J 2003, "How to Analyze Diamond Stocks", in *Resource World Magazine*, Resource World magazine Inc. Vancouver, Canada

116. Mathers, C 2004, *Crime School: Money Laundering*, Key Porter Books Limited, Toronto, Canada

117. Even-Zohar, C 2006, "Study Cites Money Laundering By U.S. Diamond Jewelry Retailers" quoting Professor J. Zdanowicz PH.D., *Diamond Intelligence Briefs*, viewed on June 12, 2006 at <http://diamondintelligence.com/magazine/magazine.asp?id=3807>

118. BBC News web site, "Guilty Plea from $200m Fraudster", viewed on December 22, 2004 at; <http://www.news.bbc.co.uk/1/hi/world/americas/1990833.stm>

119. U.S. treasury web site 2006, *Auction Goers Sure to be Pleased by These Bad Guys' Goodies*, U.S. Treasury, Seized Property Actions, viewed May 22, 2006, at <http://www.ustreas.gov/auctions/customs/p031903.html>

120. Diamond Source of Virginia web site, "Diamond Facts", viewed on August 5, 2005 at; <http//www.dsourceva.com/diamonds-facts.htm>

121. Even-Zohar, C 2005, "Dubai:Fighting Against African Corruption and Money laundering Through Rough Diamonds", *Diamond Intelligence Briefs*, Tacy Ltd., April 20, 2005

122. Tacy Ltd. Web site, 2005, *Diamond Trade Cited As Vulnerable to Money laundering by U.S. State Department Report*, Tacy Ltd. Diamond Industry Consultants, viewed May 22, 2005, at <http://www.tacyltd.com/Research_Materials_Full.asp?id=53421>

123. Robinson, J 2003, *The Sink—Crime, terror, and Dirty Money in the Off-shore World*, McClelland & Stewart, Toronto, Ontario

124. Stratem Inc. 2003, *Report on the Characteristics of the Quebec diamond market and on marketing orientation*, Stratum Inc, presented at the Matane Regional Development Aide Society, September 24, 2003.

125. Fox, W.J 2004, 'Dubai Conference Address' at the *World Diamond Council, 3rd Annual Meeting, March 30th, 2004*, Dubai—U.A.E, viewed on November 18, 2004 <http://www.fincen.gov/dubaiconferenceaddress.pdf>

126. Mathers, C 2004, *Crime School: Money Laundering*, Key Porter Books Limited, Toronto, Canada

127. Henderson, L Web site 2006, "Money laundering and Hiding Proceeds of Fraud Crimes", viewed on June 12, 2006 at <http://www.crimes-of-persuasion.com/Criminals/money_laundering.htm>

128. Even-Zohar, C 2005, "Dubai:Fighting Against African Corruption and Money laundering Through Rough Diamonds", *Diamond Intelligence Briefs*, Tacy Ltd., April 20, 2005

129. Even-Zohar, C 2005, *Terror and Diamonds in Washington*, Tacy Ltd. Diamond Industry Consultants, viewed on May 22, 2005 at <http://www.tacyltd.com/Research_Materials_Full.asp?id=54833>

130. Friscolant M, 2007, "Sorry, No Refunds on Cash Seizures", in MacLean's magazine, August 2007, Toronto.

131. Green M, 2003, "Crime of Opportunity", in *Canadian Diamonds* magazine summer 2003 issue, Up Here Publishing, Yellowknife, Canada

132. Harvey, S 2005, *Police Practice in Reducing Residential Burglary: Literature Review*, Ministry of Justice, Wellington, NZ

133. Stevenson R, and Forsythe L, 1998, "The Stolen Goods Market in New South Wales", NSW Bureau of crime Statistics and Research, NSW Attorney Generals Office, November 2, 1998.

134. Harvey, S 2005, *Police Practice in Reducing Residential Burglary: Literature Review*, Ministry of Justice, Wellington, NZ

135. Harlow G, 1998, *The Nature of Diamonds*, American Museum of natural History, Cambridge University press, Cambridge, United Kingdom.

136. Deljanin B and Sherman G 2004, *Changing the Colour of Diamonds—the High Pressure High Temperature Process Explained*, European Gemmological Laboratory U.S.A press, New York, U.S.A.

137. Canadian Guidelines with respect to the Sale and Marketing of Diamonds, Coloured Gemstones and Pearls, Revised Edition 2003, pp. 1, Established by the Jewellers Vigilance of Canada Inc.

138. Tighe, T 2004, "Compare prices before buying jewellery", in the *Edmonton Journal* March 30, 2004, viewed on October 16, 2004 at; <http://www.canada.com>

139. Robinson, J 2003, *The Sink; Crime, Terror, and Dirty Money in the Offshore World*, McClelland & Stewart Ltd., Toronto, Canada

140. Mathers, C 2004, *Crime School: Money Laundering*, Key Porter Books, Toronto, Canada

141. CBC Marketplace 1999, "Fake Diamonds", produced by Jim Nunn, aired November 30th, 1999 on Canadian Broadcasting Corporation television, Toronto, Canada

142. Robinson, J 2003, *The Sink: Crime Terror and Dirty Money in the Offshore World*, McClelland & Stewart, Toronto, Canada

143. Stevenson R, and Forsythe L, 1998, "The Stolen Goods Market in New South Wales", NSW Bureau of crime Statistics and Research, NSW Attorney Generals Office, November 2, 1998.

144. Rosen, E 1997, "Appraisal For Resale—part 2", in *Jewelers Circular Keystone Magazine,* June 1997, Reed Business Information, NY, USA

145. Cuellar, F 1998, Diamonds For Profit, ISBN 0-9668131-1-1

146. CBC Marketplace 1999, "Fake Diamonds", produced by Jim Nunn, aired November 30th, 1999 on Canadian Broadcasting Corporation television, Toronto, Canada.

147. Farah, D 1998, "Money Cleaned, Columbian Style—Contraband Used to Convert Drug Dollars", in *The Washington Post*, August 30, 1998, A22.

148. Hicks, D & Sansfacon, D 2000, *Preventing Residential Burglaries and Home Invasions*, International Center for the Prevention of crime, Belgium viewed November 28, 2004, at; <http//www.crimeprevention-intl.org>

149. Milling, T.J. 1997, "Killers, Gang Bangers and Drug Dealers go for Their Guns", at the Guns In America web site viewed May 5, 2006 at; <http://www.chron.com>

150. Hicks, D & Sansfacon, D 2000, *Preventing Residential Burglaries and Home Invasions*, International Center for the Prevention of crime, Belgium viewed November 28, 2004, at; <http//www.crimeprevention-intl.org>

151. Federal Bureau of Investigation 2004, *Crime in the United States 1999–2003, Federal Bureau of Investigation*, Washington D.C., U.S.A

152. Jewelers Security Alliance web page, viewed May 5, 2006, at; <http://www.jewelerssecurity.org>

153. Durham Regional Police web site, viewed may 5, 2006, at; <http://www.police.durham.on.ca>

154. Tacy Ltd web site 2005, Israeli Diamond Dealers Stung by US$2.5 Million Fraudster, Tacy Ltd. Diamond Industry Consultants, viewed May

22, 2005, at
<http://www.tacyltd.com/research_materials_full.asp?id=55457>

155. Federal Bureau of Investigation web site, Golden Ada Company, Federal Bureau of Investigations, Investigative Programs Organized Crime, viewed May 22, 2005, at,
<http://www.fbi.gov/hq/cid/orgcrime/csestudies/goldenada.htm>

156. Tacy Ltd. Web site, *Embracing for a Diamond Industry "Tsunami Scenario"*, Tacy Ltd. Diamond Industry Consultants, viewed on May 22, 2006, at
<http://www.tacyltd.com/Research_Materials_Full.asp?id=54909>

157. Farah, D 2004, *Blood From Stones, The Secret Financial Network of Terror*, Broadway Books, USA

158. Bell, S 2004, *Cold Terror*, John Wiley and Sons, Canada

159. Global Witness 2000, *Conflict Diamonds—Possibilities for the Identification, Certification and Control of Diamonds*, Global Witness Ltd., London, UK

160. United States Congress GOA-04-163 2003, *Terrorist Financing: U.S. Agencies Should Systematically Assess Terrorist' Use of Alternative Financing Mechanisms*, United States General Accounting Office, Washington, D.C., U.S.A

161. Robinson, J 2004, *The Sink*, McClelland & Stewart Ltd., Toronto, Canada

162. Beard, M 2002, "Industry Introduces Tanzanite Tracking", *Colored Stone*, viewed November 5th, 2004, at <http://www.colored-stone.com/stories/mar02/tanzanite.cfm>

978-0-595-46811-9
0-595-46811-X